INTEGRAL, MEASURE AND DERIVATIVE: A UNIFIED APPROACH

INTEGRAL, MEASURE AND DERIVATIVE: A UNIFIED APPROACH

G. E. SHILOV
B. L. GUREVICH

Revised English Edition
Translated and Edited by

Richard A. Silverman

DOVER PUBLICATIONS, INC.
NEW YORK

Published in Canada by General Publishing Company, Ltd., 30 Lesmill Road, Don Mills, Toronto, Ontario.
Published in the United Kingdom by Constable and Company, Ltd., 10 Orange Street, London WC2H 7EG.

This Dover edition, first published in 1977, is an unabridged and corrected republication of the English translation originally published by Prentice-Hall, Inc., in 1966.

International Standard Book Number: 0-486-63519-8
Library of Congress Catalog Card Number: 77-75774

Manufactured in the United States of America
Dover Publications, Inc.
180 Varick Street
New York, N.Y. 10014

AUTHORS' PREFACE

This volume is intended as a textbook for students of mathematics and physics, at the graduate or advanced undergraduate level. It should also be intelligible to readers with a good background in advanced calculus and sufficient "mathematical maturity."

The phrase "unified approach" in the title of the book refers to the consistent use of the Daniell scheme, which starts from the concept of an elementary integral defined (axiomatically) on a family of elementary functions. In the Introduction we explain in detail why we prefer this approach to others, in particular to the Lebesgue-Radon-Fréchet approach, which starts from axiomatic measure theory.

In preparing the American edition, we gave the book a complete overhaul. In particular, Chapter 1 was enlarged, Part 2 on the Stieltjes integral was totally rewritten, a section on Lebesgue-Stieltjes integration in infinite-dimensional spaces was added, and the order of presentation was changed in many places. We take this opportunity to thank Dr. R. A. Silverman, who while translating the book worked through all the mathematics and suggested many important improvements, resulting in a simpler treatment in some cases and a deeper one in others.

The inspiration for much of the material presented here stems from three books, all listed in the Bibliography on p. 227: Riesz and Nagy's *Functional Analysis* which treats the Daniell scheme for the case of one or several real variables, Loomis' *Introduction to Harmonic Analysis* where the Daniell scheme is presented in a general form (somewhat different from ours), and Saks' *Theory of the Integral* which gives a general method for differentiating set functions in n-space with respect to a system of cubes (the simplest example of a Vitali system). Some use has been made of text and problems borrowed from Chapter 4 and 6 of the book *Mathematical Analysis, A Special Course* (Moscow, 1961), written by the senior author. We would also like to express our gratitude to Prof. D. A. Raikov who read the entire book in manuscript and made a number of important suggestions.

G. E. S.
B. L. G.

TRANSLATOR'S PREFACE

The present book differs from most others in the same area by approaching its subject from the standpoint of the Daniell integral. Concerning the merits of this approach, I can do no better than quote from Paley and Wiener (*Fourier Transforms in the Complex Domain*, New York, 1934, p. 145): "In an ideal course on Lebesgue integration, all theorems would be developed from the point of view of the Daniell integral." As far as I know, there is no place else in the textbook literature where the Daniell scheme has been pursued with full generality, even to the point of including a complete theory of differentiation.

During the course of the translation, I had the benefit of the authors' unstinting cooperation, which I take this occasion to gratefully acknowledge. They in turn had the opportunity of examining the translation in manuscript and conferring upon it their "seal of approval." The Bibliography was prepared expressly for this edition, and is confined to books available in English. Sections marked with asterisks relate to certain side issues and can safely be omitted without loss of continuity.

R. A. S.

CONTENTS

INTRODUCTION

One of the basic concepts of analysis is that of the integral. The classical theory of integration, perfected in the middle of the last century by Cauchy and Riemann, is entirely adequate for solving many mathematical problems, both pure and applied. However, it does not meet the needs of a number of important branches of mathematics and physics of comparatively recent vintage, being deficient in at least three respects:

1. As classically defined, the integral applies only to functions of one or several variables, whereas nowadays one must be able to integrate over sets which cannot be described by a finite number of real parameters. This necessity arises, for example, in investigations ranging from probability theory and partial differential equations to hydrodynamics and quantum mechanics.
2. Even in the case of finitely many variables, only "relatively few" functions (e.g., those that are continuous, piecewise continuous or satisfy other rather strong requirements) can be integrated by using Riemann's classical definition of the integral. Some indication of the smallness of the class of Riemann-integrable functions is shown by the following fact: It is an easy matter to construct a sequence of functions $\{f_n(x)\}$ on the interval $a \leqslant x \leqslant b$, say, which satisfies the Cauchy convergence criterion "in the mean," in the sense that

$$\lim_{m,n\to\infty} \int_a^b |f_m(x) - f_n(x)|\, dx = 0,$$

without the sequence having a limit function which is Riemann integrable. This "lack of completeness" of the class of Riemann-integrable functions is a grave drawback, since completeness is well-nigh indispensable in any branch of modern analysis.

3. In the classical theory, the domain of integration X (e.g., a line or a plane) is "homogeneous" in the sense that the values of integrals over X do not change if the integrands are shifted. However, there are many problems where X can no longer be regarded as homogeneous. One can often take account of this lack of homogeneity by introducing a variable density, as is done in problems involving the vibration of inhomogeneous strings. But this device entails certain difficulties. For example, how should one define the density of a string loaded with point masses?

The above remarks amply illustrate the inadequacy of the classical theory of integration. All these difficulties disappear in the modern theory of integration, developed by some of the leading mathematicians of our time, from Lebesgue to the present day. The new theory does not require the domain X to be either finite-dimensional or homogeneous, and leads to a "sufficiently large" class of integrable functions, in particular, a class which is complete relative to convergence in the mean.

In presenting the general theory of the integral, we have chosen the Daniell method as our basic approach. This method gets to the crux of the matter more quickly and directly than the original method of Lebesgue, since it is not based on preliminary construction of a theory of measure. Moreover, from the Daniell standpoint, measure theory itself is particularly simple and natural, appearing as an almost self-evident consequence of the theory of the integral. In this regard, it should be pointed out that the Lebesgue and Daniell constructions of the integral are equivalent if finite-valued ("step") functions are chosen as the elementary functions. However, there are cases where functions other than step functions should be chosen as the elementary functions (e.g., in studying linear functionals on the space of continuous functions defined on a compact metric space), and then the Daniell method is effective while the Lebesgue method is not.

Having made these preliminary observations, we now give a brief sketch of the contents of the book. Part 1 is devoted to the integral, and consists of three chapters. In the first, we define the Riemann integral for a continuous function of n variables as the limit of a sequence of "lower Darboux sums," or, what amounts to the same thing, a nondecreasing sequence of step functions. This approach has the merit of pointing the way to further generalization, by axiomatization of certain special properties of integrals of step functions. The most basic of these properties is "upper continuity," i.e., if a nonincreasing sequence of step functions converges to zero, then so do their integrals. This generalization is carried out in Chap. 2, starting from a family of "elementary functions" defined on an arbitrary set X and equipped with an "elementary integral" which satisfies the axioms suggested by corresponding properties of integrals of step functions. The family of elementary functions is then enlarged by taking monotonic passages to the limit and forming differences. The result is a space of "summable functions," which is complete relative to the "natural" norm based on the new definition of the integral. Finally, in Chap. 3, we apply the general theory to functions of n real variables, thereby obtaining the "classical" Lebesgue integral.

In Part 2, we consider the Stieltjes integral, corresponding to the case where X is inhomogeneous. Chapter 4 is concerned with the Riemann-Stieltjes integral in n-space, constructed from a "quasi-volume" (i.e., an additive function of n-dimensional parallelepipeds, called "blocks"). Here we digress to indicate some applications of Stieltjes integration to classical analysis, based on the use of the Helly limit theorems. In Chap. 5 the Daniell scheme (described in Chap. 2) is used to construct the Lebesgue-Stieltjes integral in n-space, starting from continuous functions as elementary functions and the Riemann-Stieltjes integral as elementary integral. One can also start from step functions as elementary functions, as in the construction of the ordinary Lebesgue integral, but then an extra requirement of "upper continuity" must be imposed on the quasi-volume. However, this causes no trouble, since every quasi-volume σ of bounded variation is equivalent to an upper continuous quasi-volume $\tilde{\sigma}$, in the sense that Riemann-Stieltjes integrals of continuous functions have the same values with respect to both σ and $\tilde{\sigma}$.

In Part 3 the general Daniell scheme is used to develop a theory of measure. We start in Chap. 6 with a family of elementary functions defined on an arbitrary set X, equipped with an integral satisfying the conditions stipulated in Chap. 2. A function on X is said to be "measurable" if it is the limit of a sequence of elementary functions in the sense of convergence "almost everywhere." In particular, every summable function is measurable. A subset $E \subset X$ is said to be measurable if its characteristic function $\chi_E(x)$ is measurable and summable if $\chi_E(x)$ is summable. In the latter case, the "measure" of E is defined as the integral of $\chi_E(x)$. It follows at once from earlier considerations that measure is "countably additive." Then we give an alternative definition of the integral of a summable function $f(x)$, based on Lebesgue's original approach, in terms of the measures of the sets on which $f(x)$ takes values lying in given intervals. Chapters 7 and 8 are devoted to a deeper study of measure theory. The first of these chapters explores *constructive* measure theory, where general measurable sets are approximated by countable unions and intersections of particularly simple measurable sets (blocks in the case where X is n-space); the second deals with *axiomatic* measure theory, where a theory of the integral is constructed from a postulated "elementary measure" which is susceptible to various "extensions." Here again, consistent use of the Daniell scheme leads to great simplifications, and the two approaches, axiomatization of the integral (Chap. 2) and axiomatization of measure (Chap. 8) finally blend into a single theory. We conclude Part 3 with an introduction to the theory of Lebesgue-Stieltjes integration in infinite-dimensional spaces, a topic of great current interest.

The last part of the book (Part 4) is devoted to the theory of the derivative. In Chap. 9 we consider two countably additive set functions defined on the same abstract set X, one of which is still called a measure since it is non-negative. For the other set function, which is in general "signed" (i.e., which takes values of either sign), we establish a canonical decomposition (relative to the measure) into a discrete component and a continuous component, afterwards decomposing the continuous component in turn into a singular component and an absolutely continuous component $A(E)$. It turns out that $A(E)$ is the integral over E of a summable function $g(x)$, called the "density" of $A(E)$ (this is the celebrated Radon-Nikodým theorem). Particularizing the theory to the case of functions of one variable, we obtain the classical Lebesgue decomposition of an arbitrary (point) function of bounded variation into the sum of three terms, i.e., a discrete component, a singular component and an absolutely continuous component. The problem of finding the density $g(x)$ is examined in Chap. 10. This leads to the operation of differentiation, which we first study for the case where X is an interval $a \leqslant x \leqslant b$. We consider three different ways of defining the derivative, one based on special intervals (with binary rational end points), another on arbitrary intervals, and a third on arbitrary Borel sets. Each of these three definitions can be generalized to the case where X is an arbitrary set, equipped with a Borel measure. The first corresponds to differentiation with respect to a "net," the second to differentiation with respect to a "Vitali system," and the third to differentiation with respect to the class of all Borel subsets of X. In each case we prove that the derivative exists and equals the density $g(x)$ almost everywhere. Finally, as special cases, we prove de Possel's theorem on differentiation with respect to a net and Lebesgue's theorem on differentiation of a function of bounded variation.

Part 1

THE INTEGRAL

1

THE RIEMANN INTEGRAL AND STEP FUNCTIONS

1.1. The Riemann Integral

By an n-dimensional rectangular parallelepiped we mean a set of points $x = (x_1, \ldots, x_n)$ of the form

$$B = \{x: a_1 \leqslant x_1 \leqslant b_1, \ldots, a_n \leqslant x_n \leqslant b_n\},$$

where, naturally, it is assumed that

$$a_1 < b_1, \ldots, a_n < b_n.$$

For brevity, such parallelepipeds will henceforth be called "blocks." The largest of the numbers $b_1 - a_1, \ldots, b_n - a_n$ will be called the *size* of the block B, and the quantity

$$s(B) = (b_1 - a_1) \cdots (b_n - a_n)$$

will be called the *volume* of the block. The function $s(B)$ is an additive function of its argument, in the sense that if the block B is divided into subblocks $B_1, \ldots, B_p$ with no interior points in common (such subblocks are said to form a *partition* of B), then

$$s(B) = s(B_1) + \cdots + s(B_p).$$

A block which is fixed during the course of a given discussion will be called the "basic block," denoted by boldface $\mathbf{B}$.

We now recall how Riemann integrals are constructed. Let $f(x)$ be a bounded real function defined in a basic block $\mathbf{B}$. Let Π be a partition of $\mathbf{B}$

into subblocks $B_1, \ldots, B_p$, and in each block B_k choose an arbitrary point ξ_k $(k = 1, \ldots, p)$. Then form the *Riemann sum*

$$R_\Pi(f) = \sum_{k=1}^{p} f(\xi_k)s(B_k).$$

Let $d(\Pi)$ denote the largest size of the blocks $B_1, \ldots, B_p$, and let $\Pi_1, \ldots, \Pi_q, \ldots$ be a sequence of partitions such that $d(\Pi_q) \to 0$. If the sequence $R_{\Pi_q}(f)$ has a limit as $q \to \infty$, which is independent of the choice of the sequence Π_q [provided only that $d(\Pi_q) \to 0$] or of the points $\xi_k \in B_k$, then the limit is called the *Riemann integral* of the function $f(x)$ [over the block **B**], and we write

$$\int_{\mathbf{B}} f(x)\,dx = \lim_{d(\Pi)\to 0} R_\Pi(f). \tag{1}$$

One would now like to know the class of functions $f(x)$ for which this limit exists. In his *Cours d'Analyse* (1821), Cauchy proved that the integral (1) exists if $f(x)$ is continuous.[1] By 1837 Dirichlet had already observed that there are (discontinuous) functions which are not Riemann integrable.[2] Some time later, necessary and sufficient conditions for a function $f(x)$ to have a Riemann integral were found by Riemann, du Bois-Reymond and Lebesgue. In every case, it turns out that a Riemann-integrable function cannot be "too discontinuous." (Lebesgue's criterion for Riemann integrability will be given in Sec. 1.7.) Subsequently, various requirements of the theory led to a search for more general definitions of the integral, applicable to a much wider class of functions. The most important such definition was given by Lebesgue in 1902 (for $n = 1$), and later by Radon and Fréchet in the period 1912–1915 (for the general case). The construction of the Lebesgue integral can be approached in a variety of ways. For the reasons given in the Introduction, we choose the approach due to Daniell (1918). But first we must say more about Riemann integrals.

1.2. Lower and Upper Integrals

Let Π be a partition of the block **B** into subblocks $B_1, \ldots, B_p$, and let

$$m_k = \inf_{x\in B_k} f(x), \quad M_k = \sup_{x\in B_k} f(x) \qquad (k = 1, \ldots, p).$$

---

[1] Cauchy's proof cannot be considered rigorous, since the concept of uniform continuity was not at his disposal. The first rigorous proof of the existence of the Riemann integral of a continuous function was given by Darboux in 1875. The definition of uniform continuity and the theorem concerning the uniform continuity of a function defined on a closed interval is due to Heine (1870).

[2] For example, consider the *Dirichlet function* $\chi(x)$, $0 \leqslant x \leqslant 1$, equal to 0 for irrational x and to 1 for rational x. Then, given any partition Π, we can make $R_\Pi(\chi)$ equal to either 0 or 1, by choosing the numbers ξ_k to be either all irrational or all rational.

Then the expression

$$\underline{D}_{\Pi}(f) = \sum_{k=1}^{p} m_k s(B_k)$$

is called the *lower Darboux sum* of $f(x)$, corresponding to the partition Π. Similarly,

$$\bar{D}_{\Pi}(f) = \sum_{k=1}^{p} M_k s(B_k)$$

is called the *upper Darboux sum* of $f(x)$. Obviously, for any choice of the points $\xi_k \in B_k$ $(k = 1, \ldots, p)$, we have

$$ms(\mathbf{B}) \leqslant \underline{D}_{\Pi}(f) \leqslant R_{\Pi}(f) \leqslant \bar{D}_{\Pi}(f) \leqslant Ms(\mathbf{B}),$$

where

$$m = \inf_{x \in \mathbf{B}} f(x), \qquad M = \sup_{x \in \mathbf{B}} f(x).$$

We now compare the values of the lower and upper sums for two different partitions Π and Π' of the same basic block $\mathbf{B}$. First suppose Π' is obtained by further subdividing the blocks of the partition Π (in which case, Π' is called a *refinement* of Π). Then every term $m_k s(B_k)$ of the sum $\underline{D}_{\Pi}$ is replaced by a sum of the form

$$\sum_j m_{k_j} s(B_{k_j}),$$

where

$$B_k = \bigcup_j B_{k_j}, \qquad m_{k_j} = \inf_{x \in B_{k_j}} f(x).$$

Since $m_k \leqslant m_{k_j}$,

$$m_k s(B_k) = m_k \sum_j s(B_{k_j}) \leqslant \sum_j m_{k_j} s(B_{k_j}),$$

and hence

$$\underline{D}_{\Pi}(f) = \sum_k m_k s(B_k) \leqslant \sum_k \sum_j m_{k_j} s(B_{k_j}) = \underline{D}_{\Pi'}(f).$$

Thus, in going from the partition Π to the "finer" partition Π', the lower Darboux sum can only increase. Similarly, in going from Π to Π', the upper Darboux sum can only decrease.

Next let Π and Π' be arbitrary partitions. Then the set of all intersections of blocks of Π with blocks of Π' forms a new partition Π'', which is a refinement of both Π and Π'. But then, as just shown,

$$\underline{D}_{\Pi}(f) \leqslant \underline{D}_{\Pi''}(f) \leqslant \bar{D}_{\Pi''}(f) \leqslant \bar{D}_{\Pi'}(f), \tag{2}$$

i.e., *a lower Darboux sum can never exceed an upper Darboux sum*. Suppose we write

$$\sup_{\Pi} \underline{D}_{\Pi}(f) = \underline{\int}_{\mathbf{B}} f(x)\, dx, \qquad \inf_{\Pi} \bar{D}_{\Pi}(f) = \bar{\int}_{\mathbf{B}} f(x)\, dx,$$

where the supremum and infimum are taken with respect to *all* partitions of the block **B**. The first of these integrals is called the *lower* (*Riemann*) *integral* of $f(x)$, and the second is called the *upper integral* of $f(x)$, both over the block **B**. Then it follows from (2) that

$$\underline{\int}_{\mathbf{B}} f(x)\,dx \leqslant \overline{\int}_{\mathbf{B}} f(x)\,dx.$$

THEOREM 1. *If* $\Pi_1, \ldots, \Pi_q, \ldots$ *is a sequence of partitions of the block* **B** *such that* $d(\Pi_q) \to 0$, *then*

$$\lim_{q\to\infty} \underline{D}_{\Pi_q}(f) = \underline{\int}_{\mathbf{B}} f(x)\,dx, \tag{3}$$

and similarly,

$$\lim_{q\to\infty} \bar{D}_{\Pi_q}(f) = \overline{\int}_{\mathbf{B}} f(x)\,dx. \tag{4}$$

Proof. Given any $\varepsilon > 0$, there is a partition Π such that

$$0 \leqslant \underline{\int}_{\mathbf{B}} f(x)\,dx - \underline{D}_{\Pi}(f) < \frac{\varepsilon}{2}.$$

Consider the quantity $\underline{D}_{\Pi_q}(f)$. The blocks of the partition Π_q fall into two groups: The first group, denoted by $B_i^{(q)}$, consists of blocks which are entirely contained in blocks of Π, and the second group, denoted by $B_j^{(q)}$, consists of blocks which intersect the boundaries of blocks B_r of Π. Correspondingly, we represent $\underline{D}_{\Pi_q}$ in the form

$$\underline{D}_{\Pi_q}(f) = \sum_i m_i^{(q)} s(B_i^{(q)}) + \sum_j m_j^{(q)} s(B_j^{(q)}).$$

Let $B_i'^{(q)}$ denote the intersections of the blocks $B_j^{(q)}$ with the blocks B_r. Adding the blocks $B_i'^{(q)}$ to the blocks $B_i^{(q)}$, we obtain a new partition Π_q' of the basic block **B**, which is a refinement of the partition Π. Consequently,

$$\underline{D}_{\Pi_q}(f) = \underline{D}_{\Pi_q'} - \sum_i m_i'^{(q)} s(B_i'^{(q)}) + \sum_j m_j^{(q)} s(B_j^{(q)}), \tag{5}$$

and hence

$$\underline{\int}_{\mathbf{B}} f(x)\,dx - \frac{\varepsilon}{2} \leqslant \underline{D}_{\Pi}(f) \leqslant \underline{D}_{\Pi_q'}(f) \leqslant \underline{\int}_{\mathbf{B}} f(x)\,dx. \tag{6}$$

Now let G_Π denote the total area of the boundaries of all the blocks of the partition Π. Since the blocks $B_i'^{(q)}$ and $B_j^{(q)}$ intersect boundaries of the partition Π and have sizes no larger than $d(\Pi_q)$, each of the sums in the right-hand side of (5) is no larger than $MG_\Pi d(\Pi_q)$ in absolute value. Choosing q such that

$$MG_\Pi\, d(\Pi_q) < \frac{\varepsilon}{4},$$

we clearly have

$$0 \leqslant \underline{\int}_{\mathbf{B}} f(x)\,dx - \underline{D}_{\Pi_q}(f) \leqslant \underline{\int}_{\mathbf{B}} f(x)\,dx - \underline{D}_{\Pi'_q}(f) + \frac{\varepsilon}{2} < \varepsilon,$$

where we have used (5) and (6). This proves (3), and (4) is proved similarly.

If the Riemann integral of the function $f(x)$ exists, then the upper and lower sums must have the same limit, and hence

$$\lim_{q\to\infty} \underline{D}_{\Pi_q}(f) = \underline{\int}_{\mathbf{B}} f(x)\,dx = \int_{\mathbf{B}} f(x)\,dx = \overline{\int}_{\mathbf{B}} f(x)\,dx = \lim_{q\to\infty} \bar{D}_{\Pi_q}(f)$$

for any sequence of partitions Π_q such that $d(\Pi_q) \to 0$. Conversely, if there is at least one pair of sequences of partitions Π_q, Π'_q $(q = 1, 2, \ldots)$ with $d(\Pi_q) \to 0$, $d(\Pi'_q) \to 0$, such that

$$\lim_{q\to\infty} \underline{D}_{\Pi_q}(f) = \lim_{q\to\infty} \bar{D}_{\Pi'_q}(f),$$

then, given any sequence Π''_q with $d(\Pi''_q) \to 0$,

$$\lim_{q\to\infty} \underline{D}_{\Pi''_q}(f) = \lim_{q\to\infty} \underline{D}_{\Pi_q}(f) = \lim_{q\to\infty} \bar{D}_{\Pi'_q}(f) = \lim_{q\to\infty} \bar{D}_{\Pi''_q}(f),$$

and hence $f(x)$ is Riemann integrable.

1.3. Step Functions

Let

$$\mathbf{B} = B_1 \cup \cdots \cup B_p$$

be a partition of the basic block $\mathbf{B}$ into subblocks $B_1, \ldots, B_p$ with no interior points in common. Then a function $h(x)$ taking constant values in each of the blocks $B_1, \ldots, B_p$, i.e., such that

$$h(x) = \begin{cases} h_1 & \text{for } x \in B_1, \\ \cdot & \cdot \quad \cdot \quad \cdot \quad \cdot \\ h_p & \text{for } x \in B_p, \end{cases}$$

is called a *step function*. The function $h(x)$ can be defined in various ways (or even left undefined) on the boundary planes of the subblocks B_k, which are planes of discontinuity for $h(x)$; the values of $h(x)$ on these planes will not matter in our subsequent considerations.

The family of all step functions defined on a block $\mathbf{B}$ will be denoted by H, or if necessary, by $H(\mathbf{B})$. The set H is a linear space with the usual operations of addition and multiplication by real numbers. Thus, if $h(x)$ and $k(x)$ are step functions, so is the linear combination

$$l(x) = \alpha h(x) + \beta k(x)$$

with real coefficients α and β. In fact, if $h(x)$ is constant in the subblocks $B_1, \ldots, B_p$, while $k(x)$ is constant in the subblocks $B'_i, \ldots, B'_q$, then $l(x)$ is constant in each of the intersections

$$B_1B'_1, \ldots, B_pB'_q,$$

which together constitute a partition of $\mathbf{B}$.[3]

The space H is closed under operations other than the forming of linear combinations. For example, if $h(x)$ is a step function, so is its absolute value $|h(x)|$. Moreover, if $h(x)$ and $k(x)$ are step functions, then so are the functions

$$h_1(x) = \max\{h(x), k(x)\}, \qquad k_1(x) = \min\{h(x), k(x)\}.$$

In particular, the *positive part* $h^+(x)$ of any step function $h(x)$, defined by

$$h^+(x) = \max\{h(x), 0\}$$

is itself a step function, and so is the *negative part* $h^-(x)$, defined by

$$h^-(x) = \max\{0, -h(x)\}.$$

Next we introduce the concept of the integral of a step function $h(x)$.[4] By the *integral* of $h(x)$ over the block $\mathbf{B}$, we mean the quantity

$$Ih = \sum_{k=1}^{p} h_k s(B_k).$$

The integral of a step function has the following two properties:

a) If h, k are any two step functions and α, β are any two real numbers, then

$$I(\alpha h + \beta k) = \alpha Ih + \beta Ik.$$

b) If h and k are two step functions such that $h(x) \leqslant k(x)$ for all $x \in \mathbf{B}$, then $Ih \leqslant Ik$. In particular, if $h(x) \geqslant 0$, then $Ih \geqslant 0$.

To prove Property a, suppose $h(x)$ is constant in the blocks $B_1, \ldots, B_p$, while $k(x)$ is constant in the blocks $B'_1, \ldots, B'_q$. Then both functions are constant in the blocks $B_1B'_1, \ldots, B_pB'_q$, and moreover

$$s(B'_j) = \sum_i s(B_iB'_j), \qquad s(B_i) = \sum_j s(B_iB'_j).$$

It follows that

$$Ih = \sum_i h_i s(B_i) = \sum_i \sum_j h_i s(B_iB'_j),$$

$$Ik = \sum_j k_j s(B'_j) = \sum_j \sum_i k_j s(B_iB'_j),$$

---

[3] Naturally, any of these intersections which is empty or degenerate (with no interior points) can be omitted.

[4] This concept is distinct from the previously introduced notion of a Riemann integral, and hence will be denoted by a different symbol.

and hence

$$I(\alpha h + \beta k) = \sum_i \sum_j (\alpha h_i + \beta k_j) s(B_i B_j') = \alpha Ih + \beta Ih,$$

as asserted. Property b is proved similarly.

1.4. Sets of Measure Zero and Sets of Full Measure

In what follows, an important role will be played by coverings of sets by collections of blocks. We say that a set E (in the basic block $\mathbf{B}$) is *covered* by a collection of blocks $\{B_\alpha\}$ if every point of E is an interior point of at least one block B_α. If E is closed, we have the *finite subcovering lemma* (a variant of the familiar Heine-Borel theorem): *From every collection of blocks $\{B_\alpha\}$ covering a closed set $E \subset \mathbf{B}$, we can select a finite subcollection covering E.*

DEFINITION. *A set $Z \subset \mathbf{B}$ is called a set of measure zero if given any $\varepsilon > 0$, there exists a countable (i.e., a finite or countably infinite) subcollection of blocks $B_1, B_2, \ldots$ covering Z such that the sum of the volumes of $B_1, B_2, \ldots$ is less than ε. The empty set will also be regarded as a set of measure zero.*

Thus a *sheet*, i.e., the intersection of $\mathbf{B}$ with some hyperplane of dimension $n - 1$ parallel to a coordinate hyperplane, is a set of measure zero, since, for any $\varepsilon > 0$, there is a block $B_\varepsilon \subset \mathbf{B}$ containing the given sheet whose volume is less than ε (we need only choose B_ε to have sufficiently small thickness). On the other hand, the whole block $\mathbf{B}$ is certainly not a set of measure zero. In fact, suppose $B_1, B_2, \ldots$ is a covering of $\mathbf{B}$. By the finite subcovering lemma, we can select from $B_1, B_2, \ldots$ a finite subcollection which also covers $\mathbf{B}$. But then the sum of the volumes of even this finite number of blocks must exceed the volume $s(\mathbf{B})$, and hence cannot be less than $\varepsilon < s(\mathbf{B})$.

It is easy to see that *the union of a countable collection of sets of measure zero is itself a set of measure zero.* In fact, if $Z_1, \ldots, Z_m, \ldots$ are sets of measure zero, then, given any $\varepsilon > 0$, we cover every set Z_m by a countable collection of blocks, the sum of whose volumes is less than $\varepsilon/2^m$. As a result, the whole set $Z = Z_1 \cup Z_2 \cup \cdots$ is covered by a countable collection of blocks,[5] the sum of whose volumes is less than ε. Therefore Z is a set of measure zero, as asserted.

A set $E \subset \mathbf{B}$ is said to be a set *of full measure*, if its complement $\mathscr{C}E$ (relative to $\mathbf{B}$) is a set of measure zero. *The intersection of a countable collection*

[5] Here we use the fact that the union of countably many countable sets is countable. The sequence $Z_1, \ldots, Z_m, \ldots$ may terminate.

of sets of full measure is itself a set of full measure. In fact, if $E_1, E_2, \ldots$ are sets of full measure and if $Z_1 = \mathscr{C}E_1, Z_2 = \mathscr{C}E_2, \ldots$ are the corresponding sets of measure zero, then, as just shown, the set

$$\mathscr{C} \bigcap_m E_m = \bigcup_m \mathscr{C}E_m = \bigcup_m Z_m$$

has measure zero. Hence $\bigcap E_m$ is a set of full measure, as asserted.

If a given property holds at every point of a set of full measure in the block **B**, we say that the property holds for *almost all* points of **B** (or *almost everywhere* in **B**). There are functions which are continuous almost everywhere, i.e., continuous except on a set of measure zero. Similarly, in the class of functions that are allowed to take infinite values, there are functions which are finite almost everywhere, i.e., finite except on a set of measure zero. The set of discontinuity points of a step function has measure zero, consisting as it does of a finite number of sheets. By the same token, the set of continuity points of a step function is a set of full measure.

The following theorem can be used to give another definition of a set of measure zero, in terms of integrals of step functions:

THEOREM 2. *A set $Z \subset$ **B** is a set of measure zero if and only if, given any $\varepsilon > 0$, there exists a nondecreasing sequence of nonnegative step functions*

$$h_1^{(\varepsilon)}(x) \leqslant \cdots \leqslant h_m^{(\varepsilon)}(x) \leqslant \cdots \tag{7}$$

such that

$$Ih_m^{(\varepsilon)} < \varepsilon \qquad \textit{for every } m = 1, 2, \ldots \tag{8}$$

and

$$\sup_m h_m^{(\varepsilon)}(x) \geqslant 1 \qquad \textit{for every } x \in Z. \tag{9}$$

Proof. If Z is a set of measure zero, then, given any $\varepsilon > 0$, there exists a collection of blocks $B_1, \ldots, B_m, \ldots$ with total volume less than ε which covers the set Z. Let $h_m^{(\varepsilon)}(x)$ denote the step function which equals 1 in the blocks $B_1, \ldots, B_m$ and 0 outside these blocks. Then the sequence of step functions $h_1^{(\varepsilon)}(x), \ldots, h_m^{(\varepsilon)}(x), \ldots$ obviously satisfies (7) and (8). Moreover, any point $x_0 \in Z$ belongs to some block B_m, and hence $h_m^{(\varepsilon)}(x_0) = 1$. But this implies (9), as required.

Conversely, suppose the properties (7), (8), (9) hold, and let $B_1, \ldots, B_{r_1}$ be the collection of blocks in which the function $h_1^{(\varepsilon)}(x)$ takes values $\geqslant \frac{1}{2}$. Then the function $h_2^{(\varepsilon)}(x)$ also takes values $\geqslant \frac{1}{2}$ in the blocks $B_1, \ldots, B_{r_1}$, and in certain blocks $B_{r_1+1}, \ldots, B_{r_2}$ as well. Similarly, the function $h_3^{(\varepsilon)}(x)$ takes values $\geqslant \frac{1}{2}$ in the blocks $B_1, \ldots, B_{r_2}$, and also in certain blocks $B_{r_2+1}, \ldots, B_{r_3}$. Continuing this argument, we obtain an infinite collection of blocks $B_1, \ldots, B_{r_1}, \ldots, B_{r_p}, \ldots$, with no interior points in common. Because of (9), the set Z is contained in the union of all the blocks B_j. We now calculate the sum of the volumes of the blocks B_j. Considering only the blocks $B_1, \ldots, B_{r_m}$ in which $h_m^{(\varepsilon)}(x)$ takes values greater than $\frac{1}{2}$, and using (8), we have

$$\sum_{j=1}^{r_m} s(B_j) \leqslant 2\varepsilon.$$

If we take the limit as $m \to \infty$, this gives

$$\sum_{j=1}^{\infty} s(B_j) \leqslant 2\varepsilon.$$

The blocks B_j may not cover the set Z, since points of Z need not be interior points of the blocks B_j. However, if we replace every block B_j by a concentric block B'_j with twice the volume of B_j, we get a covering of Z by blocks B'_j with total volume $\leqslant 4\varepsilon$. Since ε is arbitrary, Z is a set of measure zero, and the proof is complete.

COROLLARY. *Given a set $Z \subset \mathbf{B}$, suppose that for every $\varepsilon > 0$, there exists a step function $h^{(\varepsilon)}(x) \geqslant 0$ such that $Ih^{(\varepsilon)}(x) < \varepsilon$ and $h^{(\varepsilon)}(x) \geqslant 1$ on Z. Then Z is a set of measure zero.*

Proof. We need merely write

$$h_1^{(\varepsilon)}(x) \equiv h_2^{(\varepsilon)}(x) \equiv \cdots \equiv h^{(\varepsilon)}(x).$$

1.5. Further Properties of Step Functions

We now prove two important lemmas:

LEMMA 1. *If a sequence of nonnegative step functions $h_1(x), \ldots, h_p(x), \ldots$ is nonincreasing,*[6] *and if $\lim_{p\to\infty} Ih_p = 0$, then*

$$\lim_{p\to\infty} h_p(x) = 0$$

almost everywhere.

Proof. The function

$$g(x) = \lim_{p\to\infty} h_p(x),$$

defined everywhere in the block $\mathbf{B}$, is nonnegative, and the set

$$G = \{x : g(x) > 0\}$$

is the union of the sequence of sets

$$G_m = \left\{x : g(x) \geqslant \frac{1}{m}\right\}.$$

[6] I.e., if $h_1(x) \geqslant \cdots \geqslant h_p(x) \geqslant \cdots$

Therefore, to show that G is a set of measure zero, it is sufficient to show that every G_m is a set of measure zero. But on every G_m we have

$$h_p(x) \geqslant g(x) \geqslant \frac{1}{m},$$

and hence

$$mh_p(x) \geqslant 1 \qquad (p = 1, 2, \ldots).$$

The function $mh_p(x)$ is a nonnegative step function and

$$I(mh_p) = mIh_p \to 0$$

as $p \to \infty$. Therefore, given any $\varepsilon > 0$, we can always find a p such that $I(mh_p) < \varepsilon$. The fact that G_m is a set of measure zero now follows from the corollary at the end of Sec. 1.4.

LEMMA 2. *If a sequence of nonnegative step functions* $h_1(x), \ldots, h_p(x), \ldots$ *is nonincreasing, and if* $\lim_{p\to\infty} h_p(x) = 0$ *almost everywhere, then*

$$\lim_{p\to\infty} Ih_p = 0.$$

Proof. First suppose $h_p(x)$ converges to zero *everywhere*, and let Z denote the set of discontinuity points of all the functions h_p. Clearly, Z is a set of measure zero. Given any $\varepsilon > 0$, we cover Z with a collection of blocks $B_1, B_2, \ldots$ whose total volume is less than ε. With each of the remaining points x' we associate an integer $m = m(x')$ such that $h_m(x') < \varepsilon$ and a block $B'(x')$ containing x' such that h_m has a constant value in $B'(x')$. Together, the blocks $B_1, B_2, \ldots$ and the blocks $B'(x')$ form a covering of the basic block $\mathbf{B}$, from which we can select a finite subcovering, whose blocks will be denoted by $B_1, \ldots, B_r, B_1', \ldots, B_q'$. Let p be the largest of the integers associated with the corresponding points $x_1', \ldots, x_q'$. Then the function $h_p(x)$ and all step functions with higher indices do not exceed ε in the blocks $B_1', \ldots, B_q'$. Moreover, in the blocks $B_1, \ldots, B_r$, whose total volume is less than ε by construction, $h_p(x)$ does not exceed M_1, the maximum of $h_1(x)$ on $\mathbf{B}$. It can be assumed that no two of blocks $B_1', \ldots, B_q'$ have interior points in common (this can always be achieved by going over to a finer collection of blocks and then excluding shared parts of blocks), and therefore the sum of the volumes of the blocks $B_1', \ldots, B_q'$ can be regarded as no larger than the volume of the basic block $\mathbf{B}$ itself. Hence, for the integral of the function $h_p(x)$ over the block $\mathbf{B}$ and for the integral of any step function with a higher index, we have the estimate

$$Ih_p \leqslant M_1\varepsilon + \varepsilon s(\mathbf{B}).$$

Since ε can be chosen arbitrarily small, it follows that $Ih_p \to 0$, as asserted.

Now suppose $h_p(x)$ does not converge to zero everywhere, but only almost everywhere. Consider the set Z of measure zero on which the sequence $h_p(x)$ fails to approach zero. According to Theorem 2, Sec. 1.4, given any $\varepsilon > 0$, there is a nonincreasing sequence of nonnegative step functions $k_p(x)$ such that

$$\sup_p k_p(x) \geqslant 1$$

for every $x \in Z$ and

$$Ik_p < \frac{\varepsilon}{M_1}$$

for every $p = 1, 2, \ldots$. Moreover, the limits

$$\lim_{p\to\infty} Ih_p \geqslant 0, \qquad \lim_{p\to\infty} Ik_p \leqslant \frac{\varepsilon}{M_1}$$

obviously exist, while the difference $h_p - M_1 k_p$ is nonincreasing and has a nonpositive limit *everywhere*. Therefore, by the first part of the proof,

$$I(h_p - M_1 k_p) \leqslant I(h_p - M_1 k_p)^+ \to 0,$$

and hence

$$\lim_{p\to\infty} Ih_p - M_1 \lim_{p\to\infty} Ik_p = \lim_{p\to\infty} I(h_p - M_1 k_p) \leqslant 0.$$

But then

$$0 \leqslant \lim_{p\to\infty} Ih_p \leqslant M_1 \lim_{p\to\infty} Ik_p \leqslant M_1 \frac{\varepsilon}{M_1} = \varepsilon.$$

Since ε is arbitrary,

$$\lim_{p\to\infty} Ih_p = 0,$$

and the lemma is proved.

1.6. Application to the Theory of the Riemann Integral

The lower Darboux sum (see Sec. 1.2)

$$\underline{D}_\Pi(f) = \sum_{k=1}^{p} m_k s(B_k)$$

can be interpreted as the integral of a "lower" step function $\underline{h}_\Pi(x)$, taking the value m_k in the block B_k. Similarly, the upper Darboux sum

$$\bar{D}_\Pi(f) = \sum_{k=1}^{p} M_k s(B_k)$$

is the integral of an "upper" step function $\bar{h}_\Pi(x)$, taking the value M_k in B_k. In this way, a sequence of partitions $\Pi_1, \ldots, \Pi_q, \ldots$ of the block **B** gives rise to a sequence of lower step functions $\underline{h}_1(x), \ldots, \underline{h}_q(x), \ldots$ and a sequence of upper step functions $\bar{h}_1(x), \ldots, \bar{h}_q(x), \ldots$ Moreover, if every partition Π_{q+1}

is a refinement of its predecessor Π_q, then the sequence of lower step functions $\underline{h}_q(x)$ is nondecreasing and the sequence of upper step functions is nonincreasing. Assuming that $d(\Pi_q) \to 0$, we introduce the *lower function*

$$\underline{f}(x) = \lim_{q\to\infty} \underline{h}_q(x)$$

and the *upper function*

$$\bar{f}(x) = \lim_{q\to\infty} \bar{h}_q(x),$$

where obviously

$$\underline{f}(x) \leqslant f(x) \leqslant \bar{f}(x).$$

THEOREM 3. *The function $f(x)$ is Riemann integrable if and only if $\underline{f}(x)$ and $\bar{f}(x)$ coincide almost everywhere.*[7]

Proof. Suppose $f(x)$ is Riemann integrable. Then

$$\lim_{q\to\infty} I\underline{h}_q = \underline{\int}_{\mathbf{B}} f(x)\,dx = \bar{\int}_{\mathbf{B}} f(x)\,dx = \lim_{q\to\infty} I\bar{h}_q,$$

and hence

$$\lim_{q\to\infty} I(\bar{h}_q - \underline{h}_q) = 0.$$

But the sequence $\bar{h}_q - \underline{h}_q$ is nonincreasing. Therefore, by Lemma 1, Sec. 1.5,

$$0 = \lim_{q\to\infty} (\bar{h}_q - \underline{h}_q) = \lim_{q\to\infty} \bar{h}_q - \lim_{q\to\infty} \underline{h}_q = \bar{f} - \underline{f}$$

almost everywhere, as asserted.

Conversely, suppose $\bar{f}(x) = \underline{f}(x)$ almost everywhere, which means that

$$\lim_{q\to\infty} (\bar{h}_q - \underline{h}_q) = 0$$

almost everywhere. Then by Lemma 2,

$$\lim_{q\to\infty} I(\bar{h}_q - \underline{h}_q) = 0,$$

and hence

$$\underline{\int}_{\mathbf{B}} f(x)\,dx = \lim_{q\to\infty} I\underline{h}_q = \lim_{q\to\infty} I\bar{h}_q = \bar{\int}_{\mathbf{B}} f(x)\,dx,$$

i.e., $f(x)$ is Riemann integrable (see Sec. 1.2), and the proof is complete.

*1.7. Invariant Definition of Lower and Upper Functions. Lebesgue's Criterion for Riemann Integrability

The definition of lower and upper functions given in the preceding section depends explicitly on the choice of the sequence of partitions Π_q. We

[7] And hence coincide almost everywhere with the function $f(x)$.

now show that lower and upper functions can be defined directly from $f(x)$, at least to within a set of measure zero. With this aim, given any $x_0 \in \mathbf{B}$, we write[8]

$$\underset{\sim}{f}(x_0) = \varliminf_{x \to x_0} f(x), \qquad \tilde{f}(x_0) = \varlimsup_{x \to x_0} f(x). \tag{10}$$

Let Π_q $(q = 1, 2, \ldots)$ be a sequence of partitions of the block $\mathbf{B}$ such that Π_{q+1} is a refinement of Π_q, and let $\underline{f}(x)$ and $\bar{f}(x)$ be the corresponding lower and upper functions. Then

$$\underline{f}(x) = \underset{\sim}{f}(x), \qquad \bar{f}(x) = \tilde{f}(x) \tag{11}$$

almost everywhere. In fact, the relations (11) hold at every point x_0 which for arbitrary q is an interior point of a block $B_q(x_0)$ of the partition Π_q. To see this, we observe that, given any $\varepsilon > 0$, there is a ball $U_\varepsilon(x_0)$ with center at the point x_0 such that $x \in U_\varepsilon(x_0)$ implies $f(x) > \underset{\sim}{f}(x_0) - \varepsilon$. The block $B_q(x_0)$ lies in the ball $U_\varepsilon(x_0)$ for sufficiently large q, and hence $f(x) > \underset{\sim}{f}(x_0) - \varepsilon$ for all $x \in B_q(x_0)$. But then

$$\underline{h}_q(x_0) = \inf_{x \in B_q(x_0)} f(x) \geqslant \underset{\sim}{f}(x_0) - \varepsilon,$$

which implies

$$\underline{f}(x_0) = \lim_{q \to \infty} \underline{h}_q(x_0) \geqslant \underset{\sim}{f}(x_0) - \varepsilon, \tag{12}$$

as well. On the other hand, the block $B_q(x_0)$ certainly contains points x such that

$$f(x) < \underset{\sim}{f}(x_0) + \varepsilon,$$

and hence

$$\underline{h}_q(x_0) = \inf_{x \in B_q(x_0)} f(x) \leqslant \underset{\sim}{f}(x_0) + \varepsilon.$$

It follows that

$$\underline{f}(x_0) = \lim_{q \to \infty} \underline{h}_q(x_0) \leqslant \underset{\sim}{f}(x_0) + \varepsilon. \tag{13}$$

Comparing (12) and (13), and taking the limit as $\varepsilon \to 0$, we find that

$$\underline{f}(x_0) = \underset{\sim}{f}(x_0).$$

In the same way, it can be shown that

$$\bar{f}(x_0) = \tilde{f}(x_0).$$

---

[8] Thus $\underset{\sim}{f}(x_0)$ is the *limit inferior* of $f(x)$ as $x \to x_0$, i.e., the smallest limiting value of $f(x)$ as $x \to x_0$. Similarly, $\tilde{f}(x_0)$ is the *limit superior* of $f(x)$ as $x \to x_0$, i.e., the largest limiting value of $f(x)$ as $x \to x_0$. [We say that c is a *limiting value* of $f(x)$ as $x \to x_0$ if, given any $\varepsilon > 0$ and any ball U with center x_0 (synonymously, any neighborhood of x_0), there is a point $x \in U$ such that $|f(x) - c| < \varepsilon$.]

THEOREM 4 (*Lebesgue's criterion for Riemann integrability*). *The function $f(x)$ is Riemann integrable if and only if the set of discontinuity points of $f(x)$ is of measure zero.*

Proof. Obviously, x_0 is a continuity point of $f(x)$ if and only if

$$\underset{\sim}{f}(x_0) = f(x_0) = \tilde{f}(x_0).$$

If $f(x)$ is Riemann integrable, then

$$\underset{\sim}{f}(x) = \underline{f}(x) = f(x) = \bar{f}(x) = \tilde{f}(x)$$

almost everywhere, and hence almost every point x is a continuity point of $f(x)$. Conversely, if the set of continuity points of $f(x)$ is of full measure, then

$$\underline{f}(x) = \underset{\sim}{f}(x) = f(x) = \tilde{f}(x) = \bar{f}(x)$$

holds on a set of full measure, and hence $f(x)$ is Riemann integrable.

1.8. Generalization of the Riemann Integral: The Key Idea

As we saw in Sec. 1.6, if a function $f(x)$ is Riemann integrable, then it is the limit (in the sense of convergence almost everywhere) of a nondecreasing sequence of step functions, in fact, of the functions $\underline{h}_q(x)$; at the same time, $f(x)$ is the limit (in the same sense) of a nonincreasing sequence of step functions, in fact, of the functions $\bar{h}_q(x)$. The converse is also true: If a function $f(x)$ is the limit (in the sense of convergence almost everywhere) of *some* nondecreasing sequence $k_q(x)$ of step functions [not necessarily of the type $\underline{h}_q(x)$], and at the same time the limit of a nonincreasing sequence of step functions $l_q(x)$, where $k_q(x) \leqslant f(x) \leqslant l_q(x)$ *everywhere*, then $f(x)$ is Riemann integrable. To see this, let Π_q denote a partition of the block $\mathbf{B}$ into subblocks in which $k_q(x)$ is constant, and construct the corresponding functions $\underline{h}_q(x)$. Then

$$k_q(x_0) \leqslant \inf_{x \in B_q(x_0)} f(x) = \underline{h}_q(x_0) \leqslant f(x_0),$$

where $B_q(x_0)$ is the block of the partition Π_q containing x_0. Since $k_q(x_0) \to f(x_0)$ and hence $\underline{h}_q(x_0) \to f(x_0)$ almost everywhere, it follows that $\underline{f}(x_0) = f(x_0)$ almost everywhere. Similarly, $\bar{f}(x_0) = f(x)$ almost everywhere, so that $f(x)$ is Riemann integrable. Here, just as in Sec. 1.6, we have

$$\int_{\mathbf{B}} f(x)\,dx = \lim_{q \to \infty} Ik_q = \lim_{q \to \infty} Il_q.$$

Now suppose that all we know about $f(x)$ is that it is the limit, in the sense of convergence almost everywhere, of a nondecreasing sequence of step functions $h_q(x)$, where the numerical sequence Ih_q has a limit (this only requires that the set of numbers $Ih_1, Ih_2, \ldots$ be bounded). Then the quantity

$$If = \lim_{q \to \infty} Ih_q$$

will be called the "integral" of f, a definition which, at the very least, does not contradict the definition of the Riemann integral for functions which are Riemann integrable. One is immediately led to ask whether the number If depends only on the function $f(x)$, since If might conceivably depend on the choice of the sequence $h_q(x)$. Not only is the answer to this question in the affirmative, but further development of the new definition leads to a theory of the integral which is free of all the difficulties discussed in the Introduction. Moreover, and this is a cardinal point, to construct the new theory we need no longer take account of the specific nature of the region $\mathbf{B}$ or of the functions $h_q(x)$, provided only that the analogues of $h_q(x)$ and their integrals have certain general properties like those already established for step functions and their integrals over a block $\mathbf{B}$ in n-dimensional space. To point up this difference, the whole construction in Chap. 2 will be carried out for functions defined on an abstract set X. In fact, we shall start from some set H of "elementary functions" $h(x)$ defined on X, assuming that the integrals Ih are already known and have certain properties, formulated as *axioms*. Then the class of integrable functions will be enlarged by using the procedure already familiar from Sec. 1.6. This whole approach lends great generality to the construction of the integral, and permits applications of the most diverse sort.

PROBLEMS

1. Let F be a closed set obtained by removing a countable collection of disjoint open intervals $\Delta_1, \ldots, \Delta_k, \ldots$ from a closed interval $[a, b]$, where the sum of the lengths of the intervals $\Delta_1, \ldots, \Delta_k, \ldots$ equals $b - a$. Show that F is of measure zero.

Hint. F is covered by the finite collection of intervals obtained by removing $\Delta_1, \ldots, \Delta_k$ from $[a, b]$.

2 (*The Cantor set*). The "middle third" of the closed interval $[0, 1]$ is removed, i.e., the open interval $(\frac{1}{3}, \frac{2}{3})$ of length $\frac{1}{3}$. Next the middle thirds of the two remaining intervals are removed, i.e., the interval $(\frac{1}{9}, \frac{2}{9})$ is removed from $[0, \frac{1}{3}]$ and $(\frac{7}{9}, \frac{8}{9})$ is removed from $[\frac{1}{3}, 1]$. Then the middle thirds of each of the four intervals $[0, \frac{1}{9}]$, $[\frac{2}{9}, \frac{1}{3}]$, $[\frac{2}{3}, \frac{7}{9}]$ and $[\frac{8}{9}, 1]$ are removed, and so on *ad infinitum*. The remaining closed set C is called the *Cantor set*. Prove that

a) C is of measure zero; b) C has the power of the continuum.

Hint. a) Use Prob. 1; b) Compare the points of C written in the ternary number system with the points of $[0, 1]$ written in the binary number system.

3. Suppose F is a closed set contained in $[a, b]$ such that the sum of the lengths of the intervals "adjacent to F" (i.e., the components of $[a, b] - F$) is *less* than $b - a$. Prove that F is not a set of measure zero.

Hint. If F were a set of measure zero, the whole interval $[a, b]$ could be covered by a finite collection of intervals with total length less than $b - a$.

2

GENERAL THEORY OF THE INTEGRAL

This chapter, in which we carry out the generalization of the integral discussed at the end of Sec. 1.8, plays a central role in the whole book. The construction given here of a space of summable functions on an arbitrary set X, equipped with a given family of elementary functions and a given elementary integral, will be the starting point for all subsequent considerations.

2.1. Elementary Functions and the Elementary Integral

Let H be a family of bounded real functions defined on a set X (these functions will henceforth be called *elementary functions*), and suppose H satisfies the following axioms:

a) H is a linear space with the usual operations of addition and multiplication by real numbers.
b) If a function $h(x)$ belongs to H, then so does its absolute value $|h(x)|$.

It follows that if $h(x)$ belongs to H, then so does its positive part $h^+(x) = \max\{h(x), 0\}$ and its negative part $h^-(x) = \max\{0, -h(x)\}$, since these functions can obviously be written as linear combinations of $h(x)$ and $|h(x)|$:

$$h^+(x) = \tfrac{1}{2}\{|h(x)| + h(x)\}, \qquad h^-(x) = \tfrac{1}{2}\{|h(x)| - h(x)\}.$$

Moreover, if two functions $h(x)$ and $k(x)$ belong to the family H, then so do the functions $\max\{h(x), k(x)\}$ and $\min\{h(x), k(x)\}$, since, as is easily verified

$$\max\{h(x), k(x)\} = [h(x) - k(x)]^+ + k(x),$$
$$\min\{h(x), k(x)\} = -\max\{-h(x), -k(x)\}.$$

Next we assume that every function $h \in H$ is assigned a real number Ih, called the *elementary integral* of h (over X), which satisfies the following axioms:

1) If h, k are any two functions in H and α, β are any two real numbers, then

$$I(\alpha h + \beta k) = \alpha Ih + \beta Ik.$$

2) *Nonnegativity axiom.* If $h(x) \geqslant 0$, then $Ih \geqslant 0$.

3) *Continuity axiom.* If $h_n(x)$ is a nonincreasing sequence of functions in H converging to zero for all $x \in X$, then $Ih_n \to 0$.

It follows from Axioms 1 and 2 that $Ih \leqslant Ik$ if $h(x) \leqslant k(x)$. In particular,

$$Ih \leqslant Ih^+ \leqslant I(|h|),$$
$$Ih \geqslant I(-|h|) = -I(|h|),$$
$$|Ih| \leqslant I(|h|)$$

for any $h \in H$.

2.2. Sets of Measure Zero and Sets of Full Measure

Of the two equivalent definitions of sets of measure zero given in Sec. 1.4, the definition patterned after Theorem 2, p. 14 is the appropriate one to follow here:

DEFINITION. *A set $Z \subset X$ is called a set of measure zero if, given any $\varepsilon > 0$, there exists a nondecreasing sequence of nonnegative functions $h_p(x) \in H$ such that $Ih_p < \varepsilon$ and*

$$\sup_p h_p(x) \geqslant 1 \text{ on } Z.$$

The empty set will also be regarded as a set of measure zero.

It is easy to see that *the union of a countable collection of sets $Z_1, \ldots, Z_n, \ldots$ of measure zero is itself a set of measure zero.* In fact, for any $\varepsilon > 0$ and n, there is a nondecreasing sequence of functions $h_p^{(n)} \in H$ $(p = 1, 2, \ldots)$ such that $Ih_p^{(n)} < \varepsilon/2^n$ and $\sup_p h_p^{(n)}(x) \geqslant 1$ on the set Z_n. But the sequence

$$h_p = \max\{h_p^{(1)}, \ldots, h_p^{(p)}\}$$

is nondecreasing, and moreover

$$Ih_p \leqslant \sum_{k=1}^{p} Ih_p^{(k)} < \varepsilon,$$

while $\sup h_p(x) \geqslant 1$ on the set Z. Therefore Z is of measure zero, as asserted.

A set $E \subset X$ is said to be a set *of full measure* if its complement (relative to X) is a set of measure zero. By taking complements, we see at once that *the intersection of a countable collection of sets of full measure is itself a set of full measure*.

As usual, if a given property holds at every point of a set of full measure, i.e., at every point of X except for a set of measure zero, then we say that the property holds for *almost all* points of X (or *almost everywhere* in X). For example, a sequence of functions $h_p(x) \in H$ converges to zero almost everywhere if there is a set E of full measure such that $h_p(x)$ converges to zero for all $x \in E$.

LEMMA. *If a nonincreasing sequence of nonnegative functions $h_p(x) \in H$ converges to zero almost everywhere, then*

$$\lim_{p\to\infty} Ih_p = 0.$$

Proof. Let

$$M_1 = \sup_{x\in X} h_1(x),$$

and let Z be the set of measure zero on which the sequence h_p does not converge to zero. Then, give any $\varepsilon > 0$, there is a nondecreasing sequence of nonnegative functions $k_p \in H$ such that $Ik_p < \varepsilon/M_1$ and $\sup\limits_p k_p(x) \geqslant 1$ on the set Z. The limits

$$\lim_{p\to\infty} Ih_p \geqslant 0, \qquad \lim_{p\to\infty} Ik_p \leqslant \frac{\varepsilon}{M_1}$$

obviously exist, while the difference $h_p - M_1k_p$ is nonincreasing and has a nonpositive limit *everywhere*. Therefore, by Axiom 3,

$$I(h_p - M_1k_p) \leqslant I(h_p - M_1k_p)^+ \to 0,$$

and hence

$$\lim_{p\to\infty} Ih_p - M_1 \lim_{p\to\infty} Ik_p = \lim_{p\to\infty} I(h_p - M_1k_p) \leqslant 0.$$

But then

$$0 \leqslant \lim_{p\to\infty} Ih_p \leqslant M_1 \lim_{p\to\infty} Ik_p \leqslant M_1 \frac{\varepsilon}{M_1} = \varepsilon.$$

Since ε is arbitrary,

$$\lim_{p\to\infty} Ih_p = 0,$$

and the lemma is proved.

If a function $h \in H$ is nonzero only on a set of measure zero, then $Ih = 0$. In fact, applying the lemma to the sequence $|h|, |h|, \ldots,$ we find that $I(|h|) = 0$ and hence

$$|Ih| \leqslant I(|h|) = 0.$$

Therefore, *if two functions $h \in H$ and $k \in H$ differ only on a set of measure zero, then $Ih = Ik$.*

The last result can be used to strengthen the lemma somewhat, i.e., the conclusion of the lemma remains true ($Ih_p \to 0$) even if the sequence h_p, which converges to zero almost everywhere, is nonincreasing only almost everywhere. In fact, replacing h_2 by $h_2' = \min(h_1, h_2)$, h_3 by $h_3' = \min(h_2', h_3)$, and so on, we alter the functions of our sequence only on a set of measure zero, which has no effect on their integrals. But then we get a sequence which is nonincreasing *everywhere* and convergent to zero almost everywhere, and the lemma applies in its original form.

The symbol $\nearrow$ will be used in connection with nondecreasing numerical sequences and also with sequences of functions which are nondecreasing on a set of full measure. Thus $h_n(x) \nearrow f(x)$ means that the sequence of functions $h_n(x)$ is nondecreasing and convergent to $f(x)$ on a set of full measure. The symbol $\searrow$ is interpreted similarly.

2.3. The Class L^+. Integration in L^+

We now introduce a class of functions, denoted by $L^+(X)$, or simply by L^+. A function $f(x)$ [which may take infinite values] is said to belong to L^+ if there exists a sequence of functions $h_n(x) \in H$ such that $h_n \nearrow f$, where the set of integrals $Ih_1, Ih_2, \ldots$ is bounded, i.e.,

$$Ih_n \leqslant C \qquad (n = 1, 2, \ldots). \tag{1}$$

First we show that every function $f(x) \in L^+$ is actually finite almost everywhere. Let $Z \subset X$ be the set of points where $f(x) = +\infty$. It can be assumed that the functions $h_n(x)$ are nonnegative, since otherwise we need only replace $h_n(x)$ by $h_n(x) - h_1(x)$. Discarding a set of measure zero, if necessary, we can assume that the sequence $h_n(x)$ is nondecreasing and convergent to $+\infty$ on the whole set Z. Given any $\varepsilon > 0$ and any $x \in Z$, the inequality

$$h_n(x) > \frac{C}{\varepsilon}$$

holds, starting from some value of n. Therefore Z is covered by the countable collection of sets

$$\left\{x : h_n(x) > \frac{C}{\varepsilon}\right\} \qquad (n = 1, 2, \ldots),$$

and hence, on the set Z we certainly have

$$\sup_n \frac{\varepsilon h_n(x)}{C} \geqslant 1.$$

At the same time, by (1),

$$I\left(\frac{\varepsilon h_n}{C}\right) = \frac{\varepsilon}{C} Ih_n \leqslant \varepsilon,$$

i.e., Z is a set of measure zero by the definition given in Sec. 2.2.

It is also apparent from the very definition of the class L^+ that if $f(x)$ belongs to L^+, then so does every function $f_1(x)$ which differs from $f(x)$ only on a set of measure zero. Obviously, every function $h(x) \in H$ belongs to L^+, and so does every function $h_1(x)$ which differs from $h(x)$ only on a set of measure zero. In particular, every function differing from zero only on a set of measure zero belongs to the class L^+.

Next we define the integral of a function f of the class L^+ by the formula

$$If = \lim_{n\to\infty} Ih_n, \tag{2}$$

where h_n is the sequence of functions of the class H figuring in the definition of the function f. Since the sequence of numbers Ih_n is nondecreasing and bounded, the limit in the right-hand side of (2) certainly exists, but we must still show that it does not depend on the choice of the sequence h_n defining the function f. This will be shown after proving the following more general fact: If h_m and k_n are two sequences of functions of the class H such that both sets $Ih_1, Ih_2, \ldots$ and $Ik_1, Ik_2, \ldots$ are bounded, and if

$$h_m \nearrow f, \qquad k_n \nearrow g, \qquad f \leqslant g$$

almost everywhere, then

$$\lim_{m\to\infty} Ih_m \leqslant \lim_{n\to\infty} Ik_n. \tag{3}$$

To see this, we hold the index m fixed and consider the nonincreasing sequence

$$h_m - k_n \qquad (n = 1, 2, \ldots)$$

of functions in H. This sequence has the limit

$$h_m - g \leqslant f - g \leqslant 0.$$

But then $(h_m - k_n)^+ \searrow 0$ (almost everywhere), and hence, by Axiom 3, $(Ih_m - k_n)^+ \searrow 0$. Since $I(h_m - k_n) \leqslant I(h_m - k_n)^+$, it follows that the integral $I(h_m - k_n) = Ih_m - Ik_n$ is nonincreasing and has a nonpositive limit, which implies

$$Ih_m \leqslant \lim_{n\to\infty} Ik_n.$$

Since this inequality holds for arbitrary m, we can take the limit as $m \to \infty$, obtaining the desired result (3).

Setting $g = f$, we find that $If \leqslant Ig$, but because of the complete equivalence of f and g, we also have $Ig \leqslant If$. It follows that $If = Ig$. Thus the definition (2) of the integral of a function $f \in L^+$ is unique. Moreover, if $f \in L^+$, $g \in L^+$, $f \leqslant g$, then $If \leqslant Ig$.

2.4. Properties of the Integral in the Class L^+

Next, by the familiar process of passing to the limit, some (but not all) of the properties of integrals in the class H can be carried over to integrals of functions in the class L^+. In fact, it is easily verified that

a) If $f \in L^+$, $g \in L^+$, then $f + g \in L^+$, and

$$I(f + g) = If + Ig.$$

b) If $f \in L^+$, then $\alpha f \in L^+$ for every $\alpha \geqslant 0$, and[1]

$$I(\alpha f) = \alpha If.$$

c) If $f \in L^+$, $g \in L^+$, then $\min(f, g) \in L^+$, $\max(f, g) \in L^+$. In particular, if $f \in L^+$, then[2]

$$f^+ = \max(f, 0) \in L^+.$$

Next we show that the class L^+ is closed under passage to the limit of nondecreasing sequences of functions with bounded integrals:

THEOREM 1. *If $f_n \in L^+$ $(n = 1, 2, \ldots)$, $f_n \nearrow f$ and $If_n \leqslant C$, then $f \in L^+$ and*

$$If = \lim_{n \to \infty} If_n.$$

Proof. For each function f_n we construct the appropriate defining sequence of functions in H:

$$\begin{array}{l} h_{11} \leqslant \cdots \leqslant h_{1n} \leqslant \cdots, \quad h_{1n} \nearrow f_1, \\ h_{21} \leqslant \cdots \leqslant h_{2n} \leqslant \cdots, \quad h_{2n} \nearrow f_2, \\ \cdots\cdots\cdots\cdots\cdots\cdots\cdots\cdots \\ h_{k1} \leqslant \cdots \leqslant h_{kn} \leqslant \cdots, \quad h_{kn} \nearrow f_k, \\ \cdots\cdots\cdots\cdots\cdots\cdots\cdots\cdots \end{array}$$

[1] Note that L^+ is not closed under subtraction or multiplication by negative numbers, since we must always deal with nondecreasing sequences of functions $h_n \in H$.

[2] The same is not true of the functions f^- and $|f|$.

Then let $h_n = \max(h_{1n}, \ldots, h_{nn})$. Obviously, h_n is also a function of the class H, and the sequence h_n is nondecreasing. Moreover

$$h_n \leqslant \max(f_1, \ldots, f_n) = f_n,$$

and hence $Ih_n \leqslant If_n \leqslant C$. Writing

$$f^* = \lim_{n \to \infty} h_n,$$

we find, by the definition of the class L^+, that $f^* \in L^+$ and

$$If^* = \lim_{n \to \infty} Ih_n.$$

But since $h_{kn} \leqslant h_n \leqslant f_n$ for any fixed k and $n \geqslant k$, passing to the limit $n \to \infty$ gives $f_k \leqslant f^* \leqslant f$. Since $f_k \nearrow f$ by hypothesis, it follows that $f^* = f$ (almost everywhere), and hence $f \in L^+$. Moreover $Ih_{kn} \leqslant Ih_n \leqslant If_n \leqslant If$, and since $Ih_n \nearrow If^* = If$, we have $If_n \nearrow If$ as well. This completes the proof.

COROLLARY. *Let $g_k \in L^+$, $g_k \geqslant 0$ be such that*

$$I\left(\sum_{k=1}^{n} g_k\right) \leqslant C \qquad (n = 1, 2, \ldots).$$

Then

$$f = \sum_{k=1}^{\infty} g_k$$

belongs to L^+, and

$$If = \sum_{k=1}^{\infty} Ig_k.$$

Proof. We need only set

$$f_n = \sum_{k=1}^{n} g_k,$$

and then apply Theorem 1.

2.5. The Class L. Integration in L

We now complete the construction of the integral by extending it from the class L^+ to a wider class $L = L(X)$, which is closed under all natural algebraic operations. By a *summable* (or *Lebesgue-integrable*) function we mean a function $\varphi(x)$ which can be represented on a set of full measure as the difference

$$\varphi = f - g$$

between two functions f and g of the class L^+. The set of all summable functions will be denoted by L. The following operations can then be carried out in L:

a) *Addition.* If $\varphi = f - g$ and $\varphi_1 = f_1 - g_1$ are summable functions, so that f, g, f_1 and g_1 belong to L^+, then

$$\varphi + \varphi_1 = (f + f_1) - (g + g_1),$$

and hence $\varphi + \varphi_1$ is summable, since $f + f_1 \in L^+$, $g + g_1 \in L^+$.

b) *Multiplication by an arbitrary real number* α. If $\alpha \geqslant 0$, then $\varphi = f - g$, $f \in L^+$, $g \in L^+$ implies $\alpha\varphi = \alpha f - \alpha g$, $\alpha f \in L^+$, $\alpha g \in L^+$, and hence $\alpha\varphi \in L$. On the other hand, $\alpha < 0$ implies $-\alpha > 0$ and then $\alpha\varphi = (-\alpha)g - (-\alpha)f$ implies $\alpha\varphi \in L$, as before. Together, a) and b) show that any linear combination of summable functions is also summable.

c) *The operations* $|\varphi|$, φ^+, φ^-. If $\varphi = f - g$, $f \in L^+$, $g \in L^+$, then $\max(f, g) \in L^+$, $\min(f, g) \in L^+$ and hence

$$|\varphi| = \max(f, g) - \min(f, g)$$

belongs to L. Since

$$\varphi^+ = \tfrac{1}{2}(|\varphi| + \varphi), \qquad \varphi^- = \tfrac{1}{2}(|\varphi| - \varphi),$$

we see that $\varphi^+ \in L$, $\varphi^- \in L$ if $\varphi \in L$. Moreover, it follows from the formulas

$$\max(\varphi, \psi) = (\varphi - \psi)^+ + \psi,$$
$$\min(\varphi, \psi) = -\max(-\varphi, -\psi)$$

that $\max(\varphi, \psi) \in L$, $\min(\varphi, \psi) \in L$ if $\varphi \in L$, $\psi \in L$.

To define the integral of a function $\varphi \in L$, suppose φ has the decomposition

$$\varphi = f - g, \quad f \in L^+, \quad g \in L^+. \tag{4}$$

Then we set

$$I\varphi = If - Ig,$$

and call $I\varphi$ the *Lebesgue integral* of the function $\varphi(x)$, conventionally written as

$$\int_X \varphi(x)\, dx.$$

It should be noted that the construction of the Lebesgue integral given here is due to Daniell,[3] and the construction originally given by Lebesgue in 1902 (with which we shall become acquainted in Part 3) is based on a different approach.

---

[3] P. J. Daniell, *A general form of integral*, Ann. of Math., **19**, 279 (1917). See also L. H. Loomis, *An Introduction to Abstract Harmonic Analysis*, D. Van Nostrand Co., Inc., Princeton, N.J. (1953).

Next we verify that the number $I\varphi$ is uniquely defined. Suppose that besides (4), there is a second decomposition

$$\varphi = f_1 - g_1, \quad f_1 \in L^+, \quad g_1 \in L^+.$$

Then we want to show that

$$If - Ig = If_1 - Ig_1, \tag{5}$$

or equivalently, that

$$If + Ig_1 = Ig + If_1. \tag{6}$$

But since $f + g_1 = g + f_1$, we have

$$I(f + g_1) = I(g + f_1),$$

because of the uniqueness of the integral in L^+, and this implies (6) and hence (5).

The integral just defined in the class L has the usual linearity properties. Let $\varphi = f - g$, $\varphi_1 = f_1 - g_1$, where f, g, f_1 and g_1 belong to the class L^+. Then

$$\varphi + \varphi_1 = (f + f_1) - (g + g_1),$$

and according to the definition,

$$\begin{aligned} I(\varphi + \varphi_1) &= I(f + f_1) - I(g + g_1) = If + If_1 - Ig - Ig_1 \\ &= (If - Ig) + (If_1 - Ig_1) = I\varphi + I\varphi_1, \end{aligned}$$

i.e., the integral of a sum equals the sum of the integrals. Moreover, if $\alpha \geqslant 0$, then

$$\begin{aligned} I(\alpha\varphi) = I(\alpha f - \alpha g) = I(\alpha f) - I(\alpha g) &= \alpha If - \alpha Ig \\ &= \alpha(If - Ig) = \alpha I\varphi. \end{aligned}$$

On the other hand,

$$I(-\varphi) = I(g - f) = Ig - If = -I\varphi,$$

and hence, if $\alpha < 0$, we have

$$I(\alpha\varphi) = I(-|\alpha|\,\varphi) = -I(|\alpha|\,\varphi) = -|\alpha|\,I\varphi = \alpha I\varphi.$$

Therefore, regardless of its sign, the number α can be brought in front of the integral sign.

Next we note that if $\varphi \in L$, $\varphi \geqslant 0$, then $I\varphi \geqslant 0$. In fact, if $\varphi = f - g$, $f \in L^+$, $g \in L^+$ and $\varphi \geqslant 0$, then $f \geqslant g$ and hence $If \geqslant Ig$, or equivalently, $I\varphi = If - Ig \geqslant 0$. Similarly $\varphi_1 \in L$, $\varphi_2 \in L$, $\varphi_1 \leqslant \varphi_2$ implies

$$I\varphi_1 \leqslant I\varphi_2,$$

and hence

$$|I\varphi| \leqslant I(|\varphi|)$$

since $\pm\varphi \leqslant |\varphi|$.

Remark. Other conditions can be imposed on the functions f and g figuring in the decomposition $\varphi = f - g$, $f \in L^+$, $g \in L^+$ of a summable function. For example, given any $\varepsilon > 0$, we can always choose g such that $g \geqslant 0$, $Ig < \varepsilon$. To see this, consider a sequence $h_n \nearrow g$ of functions in H such that $Ig = \lim_{n\to\infty} Ih_n$, and then write

$$\varphi = f - g = (f - h_n) - (g - h_n) = f_n - g_n.$$

Here f_n belongs to L^+, since $f_n = f - h_n = f + (-h_n)$ is a sum of two functions in L^+ (the second function even belongs to H), and the same is true of g_n. It is obvious that for sufficiently large n, the function $g_n = g - h_n$ satisfies the stipulated conditions $g_n \geqslant 0$, $Ig_n < \varepsilon$. In fact, if $\varphi \geqslant 0$, the function $f_n = f - h_n \geqslant f - g = \varphi$ also turns out to be nonnegative.

2.6. Levi's Theorem

We now prove an important theorem on term-by-term integration of series with nonnegative terms:

THEOREM 2 (*Levi's theorem*). *Let* $\varphi_k \in L$, $\varphi_k \geqslant 0$ *be such that*

$$I\left(\sum_{k=1}^{n} \varphi_k\right) \leqslant C \qquad (n = 1, 2, \ldots).$$

Then

$$\varphi = \sum_{k=1}^{\infty} \varphi_k$$

is a summable function, and

$$I\varphi = \sum_{k=1}^{\infty} I\varphi_k.$$

Proof. Using the remark at the end of Sec. 2.5, we represent each φ_k in the form

$$\varphi_k = f_k - g_k, \quad f_k \in L^+, \quad g_k \in L^+,$$

where $f_k \geqslant 0$, $g_k \geqslant 0$, $Ig_k < 1/2^k$ $(k = 1, 2, \ldots)$. Then the functions g_k satisfy all the conditions of the corollary to Theorem 1, since $g_k \geqslant 0$ and

$$I\left(\sum_{k=1}^{n} g_k\right) \leqslant 1.$$

Therefore

$$g = \sum_{k=1}^{\infty} g_k$$

belongs to L^+, and

$$Ig = \sum_{k=1}^{\infty} Ig_k.$$

Moreover, the functions f_k also satisfy the conditions of the corollary. In fact, $f_k \geqslant 0$ and

$$I\left(\sum_{k=1}^{n} f_k\right) = I\left(\sum_{k=1}^{n} \varphi_k\right) + I\left(\sum_{k=1}^{n} g_k\right) \leqslant C + 1.$$

Therefore

$$f = \sum_{k=1}^{\infty} f_k$$

also belongs to L^+, and

$$If = \sum_{k=1}^{\infty} If_k.$$

It follows that

$$\varphi = \sum_{k=1}^{\infty} \varphi_k = \sum_{k=1}^{\infty} f_k - \sum_{k=1}^{\infty} g_k = f - g$$

belongs to L, and

$$I\varphi = If - Ig = \sum_{k=1}^{\infty} f_k - \sum_{k=1}^{\infty} g_k = \sum_{k=1}^{\infty} I(f_k - g_k) = \sum_{k=1}^{\infty} I\varphi_k,$$

as asserted.

COROLLARY 1. *If $\psi_n \in L$ $(n = 1, 2, \ldots)$, $\psi_n \nearrow \psi$ and $I\psi_n \leqslant C$, then $\psi \in L$ and*

$$I\psi = \lim_{n\to\infty} I\psi_n.$$

Proof. We need only set

$$\varphi_1 = \psi_1, \quad \varphi_2 = \psi_2 - \psi_1, \ldots, \quad \varphi_{n+1} = \psi_{n+1} - \psi_n, \ldots,$$

and then apply Levi's theorem. Of course, a similar result also holds for a nonincreasing sequence $\psi_n \searrow \psi$, provided that $I\psi_n \geqslant C$.

It is clear that if a function $\varphi(x) \in L$ is nonzero only on a set of measure zero, then $I\varphi = 0$. We now ask whether the converse is true, i.e., does $I\varphi = 0$ imply that $\varphi(x) = 0$ almost everywhere? Naturally, it must now be assumed that φ does not change sign (e.g., is nonnegative), since otherwise $I\varphi$ can vanish because of mutual cancellation of the integrals $I\varphi^+$ and $I\varphi^-$. Thus, assuming that $\varphi_0 \in L$, $\varphi_0 \geqslant 0$ and $I\varphi_0 = 0$, we now show that $\varphi_0 = 0$ almost everywhere. Let $\varphi_n = n\varphi_0$. Then the functions φ_n converge to a limit φ equal to zero where φ_0 vanishes and to $+\infty$ where $\varphi_0 > 0$. According to Corollary 1, the limit function φ must be summable. But then $\varphi(x) = +\infty$ only on a set of measure zero, and hence $\varphi_0(x) > 0$ only on a set of measure zero. This proves

COROLLARY 2. *If the integral of a nonnegative summable function $\varphi_0(x)$ is zero, then $\varphi_0(x) = 0$ almost everywhere.*

COROLLARY 3. *Given a set $Z \subset X$, suppose that for every $\varepsilon > 0$, there exists a sequence of summable functions*

$$0 \leqslant \varphi_1^{(\varepsilon)}(x) \leqslant \cdots \leqslant \varphi_n^{(\varepsilon)}(x) \leqslant \cdots$$

such that

$$I\varphi_n^{(\varepsilon)} < \varepsilon \qquad (n = 1, 2, \ldots)$$

and

$$\sup_n \varphi_n^{(\varepsilon)}(x) \geqslant 1 \qquad (x \in Z).$$

Then Z is a set of measure zero.

Proof. If the $\varphi_n^{(\varepsilon)}(x)$ are elementary functions, the corollary follows at once from the definition of a set of measure zero. In the general case, let

$$\varphi^{(\varepsilon)}(x) = \lim_{n\to\infty} \varphi_n^{(\varepsilon)}(x).$$

Then, by Corollary 1, $\varphi^{(\varepsilon)}(x)$ is summable and

$$I\varphi^{(\varepsilon)} = \lim_{n\to\infty} I\varphi_n^{(\varepsilon)} \leqslant \varepsilon.$$

Now choosing $\varepsilon = 1, 1/2, \ldots, 1/n, \ldots$, let

$$\psi_1 = \varphi^{(1)},\ \psi_2 = \min\{\varphi^{(1)}, \varphi^{(1/2)}\}, \ldots, \psi_n = \min\{\varphi^{(1)}, \ldots, \varphi^{(1/n)}\}, \ldots$$

The functions ψ_n are nonnegative, and $\geqslant 1$ on the set Z. Moreover

$$\psi_1(x) \geqslant \cdots \geqslant \psi_n(x) \geqslant \cdots$$

and

$$I\psi_n \leqslant I\varphi^{(1/n)} \leqslant \frac{1}{n}.$$

If

$$\psi(x) = \lim_{n\to\infty} \psi_n(x),$$

then, by Corollary 1, $\psi \in L$ and

$$I\psi = \lim_{n\to\infty} I\psi_n = 0.$$

Clearly $\psi(x)$ is nonnegative, and $\geqslant 1$ on the set Z. According to Corollary 2, the set

$$Z_1 = \{x\colon \psi(x) > 0\},$$

which obviously contains Z, is of measure zero. But then Z itself is of measure zero, as asserted.

2.7. Lebesgue's Theorem

From now on, we shall consider arbitrary (nonmonotonic) passages to the limit. Classical examples show that we cannot expect theorems of the form "$\varphi_n \to \varphi$ *implies* $I\varphi_n \to I\varphi$" to hold without further assumptions about the way the sequence φ_n converges to its limit. For example, consider the functions

$$\varphi_n(x) = \begin{cases} n \sin nx & \text{for } 0 \leqslant x \leqslant \dfrac{\pi}{n}, \\ 0 & \text{for } \dfrac{\pi}{n} < x \leqslant \pi. \end{cases}$$

Then the sequence φ_n converges to zero for every $x \in [0, \pi]$, but at the same time, $I\varphi_n$ does not converge to $I\varphi$ (in fact, $I\varphi_n = 2$ for every n).

Let $L(\varphi_0)$ denote the set of all summable functions φ which satisfy (almost everywhere) the inequality

$$-\varphi_0 \leqslant \varphi \leqslant \varphi_0, \tag{7}$$

where φ_0 is a fixed nonnegative summable function. Obviously,

$$-I\varphi_0 \leqslant I\varphi \leqslant I\varphi_0$$

for every function $\varphi \in L(\varphi_0)$. Moreover, the limit φ of a monotonic sequence of functions in $L(\varphi_0)$, whether decreasing or increasing, will clearly satisfy the inequality (7) on some set of full measure (just like the φ_n themselves), and as we have seen above, φ is also summable. In other words, the set $L(\varphi_0)$ is closed under monotonic passages to the limit. It should also be noted that given *any* sequence $\varphi_n \in L(\varphi_0)$, we can assert that the functions

$$\sup \{\varphi_1(x), \ldots, \varphi_n(x), \ldots\} \tag{8}$$

and

$$\inf \{\varphi_1(x), \ldots, \varphi_n(x), \ldots\} \tag{9}$$

also belong to $L(\varphi_0)$. In fact, (8) is the limit as $n \to \infty$ of the nondecreasing sequence

$$\max \{\varphi_1(x), \ldots, \varphi_n(x)\} \in L(\varphi_0),$$

while (9) is the limit of the nonincreasing sequence

$$\min \{\varphi_1(x), \ldots, \varphi_n(x) \in L(\varphi_0).$$

Now let $\varphi_n \in L(\varphi_0)$ be any sequence converging almost everywhere to a function φ. Then φ also belongs to $L(\varphi_0)$. To see this, we need only show that φ can be represented as the limit of a monotonic sequence of functions of the class $L(\varphi_0)$. As just shown, the functions

$$\psi_n(x) = \sup \{\varphi_n(x), \varphi_{n+1}(x), \ldots\}$$
$$\chi_n(x) = \inf \{\varphi_n(x), \varphi_{n+1}(x), \ldots\}$$

are summable and belong to $L(\varphi_0)$. By considering only values of x where the functions $\varphi_n(x)$ converge to $\varphi(x)$, we see that

$$\psi_n(x) \geqslant \lim_{p \to \infty} \varphi_{n+p}(x) = \varphi(x),$$
$$\chi_n(x) \leqslant \lim_{p \to \infty} \varphi_{n+p}(x) = \varphi(x)$$

for almost all x. Moreover, the least upper bound of the set $\varphi_n, \varphi_{n+1}, \ldots$ can only decrease if the first function φ_n is omitted, while the greatest lower bound can only increase. Therefore

$$\psi_{n+1}(x) \leqslant \psi_n(x),$$
$$\chi_{n+1}(x) \geqslant \chi_n(x),$$

i.e., the sequence $\psi_n(x)$ is nonincreasing and the sequence $\chi_n(x)$ is nondecreasing. But then $\varphi_n(x) \to \varphi(x)$ implies $\psi_n(x) \searrow \psi(x)$ and $\chi_n(x) \nearrow \chi(x)$, and hence $\varphi(x)$ is the limit of a nondecreasing sequence of functions of the class $L(\varphi_0)$ [and at the same time, the limit of a nonincreasing sequence of functions of this class]. It follows that $\varphi \in L(\varphi_0)$, as asserted, and moreover

$$I\chi_n \nearrow I\varphi, \quad I\psi_n \searrow I\varphi, \quad I\chi_n \leqslant I\varphi_n \leqslant I\psi_n,$$

so that $I\varphi_n \to I\varphi$. Thus we have proved

THEOREM 3 (*Lebesgue's theorem*).[4] *If a sequence of summable functions φ_n converges almost everywhere to a function φ and satisfies the condition*

$$|\varphi_n(x)| \leqslant \varphi_0(x) \in L \qquad (n = 1, 2, \ldots), \tag{10}$$

then φ is summable and

$$I\varphi = \lim_{n\to\infty} I\varphi_n. \tag{11}$$

2.8. Summability of Almost-Everywhere Limits

In some cases where the condition $\varphi_n \to \varphi$ does not imply $I\varphi_n \to I\varphi$, we can still draw conclusions about the summability of φ and deduce an estimate for $I\varphi$.

2.8.1. Measurable functions. For example, if instead of the condition (10) figuring in Lebesgue's theorem, it is assumed only that $\varphi_n(x) \in L$, $\varphi_n(x) \to \varphi(x)$ and

$$|\varphi(x)| \leqslant \varphi_0(x) \in L, \tag{12}$$

then $\varphi(x)$ is summable, but we no longer have the limit relation (11). In fact,

$$\varphi(x) = \lim_{n\to\infty} \psi_n(x),$$

where $\psi_n(x)$ is the function $\varphi_n(x)$ truncated from above at the level $\varphi_0(x)$ and from below at the level $-\varphi_0(x)$, i.e.,

$$\psi_n(x) = \max\{\min[\varphi_n(x), \varphi_0(x)], -\varphi_0(x)\}.$$

---

[4] Often called *Lebesgue's theorem on term-by-term integration*, or *Lebesgue's bounded convergence theorem.*

Obviously, $\psi_n(x)$ is summable and $|\psi_n(x)| \leqslant \varphi_0(x)$. Therefore, according to Lebesgue's theorem, $\varphi(x)$ is summable and we have the estimate

$$|I\varphi| \leqslant I\varphi_0.$$

In this connection, we introduce the following important concept: If a sequence of elementary functions converges almost everywhere to a function $\varphi(x)$, then $\varphi(x)$ is said to be *measurable*. For the time being, we point out only a few simple properties of measurable functions.[5] By their very construction, all summable functions are measurable, but the class of summable functions is in general only a proper subset of the class of measurable functions. The inequality (12) gives a simple condition guaranteeing the summability of a measurable function $\varphi(x)$, i.e., every measurable function whose absolute value is bounded by a summable function is itself summable. To see this, we need only note that

$$\varphi(x) = \lim_{n\to\infty} h_n(x),$$

where the functions $h_n(x)$ are elementary and hence summable, and then apply the result just proved.

2.8.2. Fatou's lemma. We can replace the condition (10) figuring in Lebesgue's theorem by the weaker condition

$$I(|\varphi_n|) \leqslant C.$$

Then the limit function

$$\varphi = \lim_{n\to\infty} \varphi_n$$

is again summable, but the limit relation (11) is replaced by the estimate

$$I(|\varphi|) \leqslant C.$$

We begin by proving this for the case where the functions φ_n are nonnegative:

LEMMA (*Fatou's lemma*). *Let $\varphi_n \in L$, $\varphi_n \geqslant 0$ be such that $\varphi_n \to \varphi$ almost everywhere and $I\varphi_n \leqslant C$. Then φ is summable, and*

$$0 \leqslant I\varphi \leqslant C.$$

Proof. If

$$\chi_n = \inf\{\varphi_n, \varphi_{n+1}, \ldots\} \geqslant 0,$$

then, as before, the functions χ_n form a nondecreasing sequence converging almost everywhere to φ. Moreover $\chi_n \leqslant \varphi_n$, $I\chi_n \leqslant I\varphi_n \leqslant C$, and hence by Corollary 1, p. 33, the function φ is summable and $I\chi_n \nearrow I\varphi$. In particular,

$$0 \leqslant I\varphi = \lim_{n\to\infty} I\chi_n \leqslant C,$$

as asserted.

[5] Part 3 will be largely devoted to a study of measurable functions.

Returning to the general case $\varphi_n(x) \to \varphi(x)$, $I(|\varphi_n|) \leqslant C$, we note that by Fatou's lemma, $|\varphi(x)|$ is summable and $I(|\varphi|) \leqslant C$. But then, by the result of Sec. 2.8.1, $\varphi(x)$ itself is summable, as required.

2.9. Completeness of the Space L. The Riesz-Fischer Theorem

We begin by recalling the definition of a *normed linear space*. A linear space R consisting of elements $\varphi, \psi, \ldots$ is said to be *normed* if with every element $\varphi \in R$ there is associated a nonnegative number $\|\varphi\|$, called the *norm* of φ, which has the following properties:

a) $\|\varphi\| > 0$ if $\varphi \neq 0$, and $\|0\| = 0$.
b) $\|\alpha\varphi\| = |\alpha| \, \|\varphi\|$ for every $\varphi \in R$ and every real number α.
c) *The triangle inequality*. $\|\varphi + \psi\| \leqslant \|\varphi\| + \|\psi\|$ for every $\varphi \in R$, $\psi \in R$.

Given a sequence of elements $\varphi_n \in R$, we say that φ is the *limit* of φ_n if

$$\lim_{n \to \infty} \|\varphi - \varphi_n\| = 0.$$

A sequence φ_n is said to be a *Cauchy* (or *fundamental*) *sequence* if

$$\lim_{m,n \to \infty} \|\varphi_m - \varphi_n\| = 0,$$

i.e., if given any $\varepsilon > 0$, there exists an integer N such that $\|\varphi_m - \varphi_n\| < \varepsilon$ whenever $m > N$, $n > N$. A normed linear space R is said to be *complete* if it satisfies the *Cauchy criterion*, i.e., if every Cauchy sequence $\varphi_n \in R$ has a limit $\varphi \in R$ (relative to the norm $\| \; \|$).

Next we construct a normed linear space of summable functions φ, by equipping L (which is already a normed linear space) with the norm

$$\|\varphi\| = I(|\varphi|). \tag{13}$$

The fact that (13) satisfies Properties b and c follows at once from the basic properties of the integral. Strictly speaking, this choice of norm does not satisfy Property a, since $I(|\varphi|) = 0$ does not imply that φ vanishes identically. However, this discrepancy is easily removed by the simple device of identifying all elements of L which differ only on a set of measure zero,[6] since, according to Corollary 2, p. 33, $I(|\varphi|) = 0$ implies that $\varphi = 0$ almost everywhere.

The effort we have expended in constructing the space L is to a large extent justified by the following important result:

---

[6] More precisely, our normed space is not L but another space whose elements are classes of functions of L differing only on sets of measure zero. For example, the zero element of this new space is the class consisting of all functions which are zero almost everywhere. We shall use the same symbol L to denote both spaces.

THEOREM 4 (*Riesz-Fischer theorem*). *The space L, equipped with the norm* (13), *is complete, i.e., every Cauchy sequence* φ_n *of summable functions has a summable limit* (*in the L-norm*).

Proof. It is enough to show that some subsequence φ_{n_k} of the Cauchy sequence φ_n has a limit $\varphi \in L$, since then φ will also be the limit of the whole sequence φ_n. This follows from the inequality

$$\|\varphi - \varphi_n\| \leqslant \|\varphi - \varphi_{n_k}\| + \|\varphi_{n_k} - \varphi_n\|$$

and the fact that the second term on the right goes to zero as $n \to \infty$ and $n_k \to \infty$ (since φ_n is a Cauchy sequence). Clearly, we can always find an increasing sequence of indices n_k such that

$$\|\varphi_n - \varphi_{n_k}\| < \frac{1}{2^k} \qquad (k = 1, 2, \ldots)$$

for $n > n_k$. In particular,

$$\|\varphi_{n_{k+1}} - \varphi_{n_k}\| < \frac{1}{2^k},$$

which means that

$$I(|\varphi_{n_{k+1}} - \varphi_{n_k}|) < \frac{1}{2^k}.$$

But then, according to Levi's theorem, the series of summable functions

$$\sum_{k=1}^{\infty} |\varphi_{n_{k+1}} - \varphi_{n_k}|$$

converges almost everywhere, and hence the same is true of the series

$$\sum_{k=1}^{\infty} (\varphi_{n_{k+1}} - \varphi_{n_k}),$$

with partial sums

$$\sum_{k=1}^{N} (\varphi_{n_{k+1}} - \varphi_{n_k}) = \varphi_{n_{N+1}} - \varphi_{n_1}.$$

This means that the sequence φ_{n_k} has a limit (almost everywhere) as $k \to \infty$. Let φ denote this limit. Then, for fixed k, the function $\varphi_{n_p} - \varphi_{n_k}$ approaches $\varphi - \varphi_{n_k}$ almost everywhere as $p \to \infty$. Since

$$I(|\varphi_{n_p} - \varphi_{n_k}|) = \|\varphi_{n_p} - \varphi_{n_k}\| < \frac{1}{2^k} \qquad (p > k),$$

it follows from the result of Sec. 2.8.2 that $\varphi - \varphi_{n_k}$ is summable, and hence φ itself is summable. Moreover, by the same result,

$$\|\varphi - \varphi_{n_k}\| = I(|\varphi - \varphi_{n_k}|) \leqslant \frac{1}{2^k}.$$

Therefore φ_{n_k} converges to φ in the norm of the space L, and the proof is complete.

Finally, we prove that the set H of all elementary functions is dense in the space L. Since every function in L is the difference of two functions in L^+, we need only verify that every function $f \in L^+$ is the limit (in the norm of L) of a sequence of functions $h_n \in H$. The natural choice for this sequence is the sequence defining f. Then $h_n \nearrow f$, $Ih_n \nearrow If$ and

$$\|f - h_n\| = I(f - h_n) = If - Ih_n \to 0,$$

as required.

2.10. Fubini's Theorem

Next we consider integration over the product of two sets, deriving the formula for reduction of a double integral to an iterated integral. It will be recalled from calculus that the double Riemann integral of a continuous function $f(x, y)$ can be expressed in terms of two single Riemann integrals by using the rule

$$\iint\limits_{\substack{a_1 \leqslant x_1 \leqslant b_1 \\ a_2 \leqslant x_2 \leqslant b_2}} f(x, y)\,dx\,dy = \int_{a_2}^{b_2} \left\{ \int_{a_1}^{b_1} f(x, y)\,dx \right\} dy.$$

As we now show, there is a similar rule in the general theory of integration:

THEOREM 5 (*Fubini's theorem*). *Given two sets X and Y, with Cartesian product $W = X \times Y$,*[7] *let $L(X)$, $L(Y)$ and $L(W)$ be corresponding spaces of summable functions, equipped with integrals I_X, I_Y and $I_W = I$. Suppose the family $H(W)$ of elementary functions generating $L(W)$ has the following properties:*[8]

a) *Every function $h(x, y) \in H(W)$ is summable in x for almost all y;*
b) *The integral $I_X h(x, y)$ is summable in y;*
c) $Ih = I_Y\{I_X h(x, y)\}$.

Then $L(W)$ has the same properties, i.e., every function $\varphi(x, y) \in L(W)$ is summable in x for almost all y, the integral $I_X \varphi(x, y)$ is summable in y, and

$$I\varphi = I_Y\{I_X \varphi(x, y)\}.$$

Proof. Let Φ denote the set of all functions $\varphi \in L(W)$ for which the theorem holds. By hypothesis, Φ contains all the elementary functions $h(x, y)$. The theorem will be proved once we succeed in showing that *every* function in $L(W)$ belongs to Φ. This will be established in five steps.

---

[7] By $X \times Y$ is meant the set of all ordered pairs (x, y), where $x \in X$ and $y \in Y$.

[8] In addition to the usual properties of a family of elementary functions.

Step 1. Obviously, Φ *is closed under the formation of linear combinations*, i.e., if $\varphi_1 \in \Phi$, $\varphi_2 \in \Phi$, then $\alpha_1\varphi_1 + \alpha_2\varphi_2 \in \Phi$, where α_1, α_2 are arbitrary real numbers.

Step 2. Φ *is closed under monotonic passages to the limit.* More exactly, let $\varphi_1(x, y), \varphi_2(x, y), \ldots$ be a sequence of functions in Φ which is (everywhere) monotonic, and suppose the integrals $I\varphi_n$ form a bounded (numerical) sequence. Then

$$\varphi(x, y) = \lim_{n\to\infty} \varphi_n(x, y)$$

belongs to Φ. For example, suppose the sequence $\varphi_n(x, y)$ is nondecreasing, and let $g_n(y) = I_X\varphi_n(x, y)$. Then the sequence $g_n(y)$ is also nondecreasing, and the integrals $I_Y g_n$ form a bounded sequence:

$$I_Y g_n = I_Y\{I_X\varphi_n(x, y)\} = I\varphi_n \nearrow I\varphi.$$

By Corollary 1 to Levi's theorem, $g_n(y)$ converges to a summable function $g(y)$, which must be finite almost everywhere, and moreover

$$I_Y g = \lim_{n\to\infty} I_Y g_n = I\varphi.$$

Let $E \subset Y$ be the set of full measure on which the function $g(y)$ is finite, and let y be a (temporarily) fixed point of E. Then the sequence $\varphi_n(x, y)$ is nondecreasing and the integrals $I_X\varphi_n(x, y)$ form a bounded sequence:

$$I_X\varphi_n(x, y) = g_n(y) \nearrow g(y).$$

Therefore, applying Corollary 1 again, we find that the limit function $\varphi(x, y)$ is summable in x (for the given value of y), and

$$\lim_{n\to\infty} I_X\varphi_n(x, y) = g(y) = I_X\varphi(x, y).$$

But then

$$I\varphi = I_Y g(y) = I_Y\{I_X\varphi(x, y)\},$$

and hence $\varphi(x, y) \in \Phi$, as asserted.

Step 3. Φ *contains every function* $z(x, y)$ *which is different from zero only on a set* $Z \subset W$ *of measure zero.* First suppose the values taken by $z(x, y)$ on Z lie between 0 and 1. Since Z is a set of measure zero, given any $m = 1, 2, \ldots$, we can construct a nondecreasing sequence of nonnegative elementary functions $h_n^{(m)}(x, y)$ such that

$$Ih_n^{(m)}(x, y) < \frac{1}{m}, \qquad \lim_{n\to\infty} h_n^{(m)}(x, y) \geqslant 1 \text{ on Z}.$$

Moreover, it can always be assumed that

$$h_n^{(m+1)}(x, y) \leqslant h_n^{(m)}(x, y),$$

since otherwise we can replace h_n^{m+1} by $\min\{h_n^{m+1}(x, y),\ h_n^{(m)}(x, y)\}$. The function

$$h^{(m)}(x, y) = \lim_{n\to\infty} h_n^{(m)}(x, y)$$

is a monotonic limit of (elementary) functions of Φ, and hence belongs to Φ, by Step 2. For the same reason, the function

$$h(x, y) = \lim_{m\to\infty} h^{(m)}(x, y)$$

also belongs to Φ, and clearly

$$Ih^{(m)} = \lim_{n\to\infty} Ih_n^{(m)} \leqslant \frac{1}{m},$$

$$Ih = \lim_{m\to\infty} Ih^{(m)} = 0.$$

Moreover $h(x, y) \geqslant z(x, y)$ on Z, since $h^{(m)}(x, y) \geqslant z(x, y)$ on Z for all $m = 1, 2, \ldots$ If $g(y) = I_X h(x, y)$, then

$$I_Y g(y) = I_Y I_X h(x, y) = Ih = 0,$$

by Step 2, and hence $g(y) = 0$ for almost all y (by Corollary 2 to Levi's theorem). But then, for these values of y, the function $h(x, y)$ vanishes for almost all x, and hence the same is true of the function $z(x, y)$. It follows that $I_X z(x, y) = 0$, and hence

$$Iz(x, y) = 0 = I_Y\{I_X z(x, y)\},$$

i.e., $z(x, y) \in \Phi$, as asserted.

If $z(x, y) \geqslant 0$ is an arbitrary function vanishing outside the set Z, then, introducing the function $l(x, y)$ equal to 1 on Z and 0 outside Z, we have

$$z(x, y) = \lim_{n\to\infty} n \min\left\{l(x, y), \frac{1}{n} z(x, y)\right\}.$$

Therefore $z(x, y) \in \Phi$, by the argument just given, together with Step 2. The general case, where $z(x, y)$ can have either sign, reduces to the case just considered by writing $z = z^+ - z^-$.

Step 4. Φ contains every function $f(x, y) \in L^+(W)$. By the definition of $L^+(W)$, there is a sequence of elementary functions $h_n(x, y)$ such that

$$h_n(x, y) \nearrow f(x, y), \qquad Ih_n \nearrow If,$$

where the sequence $h_n(x, y)$ need only be nondecreasing almost everywhere. Let $\hat{f}(x, y)$ be the limit of the sequence

$$\hat{h}_1(x, y) = h(x, y), \qquad \hat{h}_2(x, y) = \max\{h_1(x, y), h_2(x, y)\}, \ldots,$$

which is nondecreasing *everywhere*. Note that the functions $\hat{h}_n$ and h_n coincide almost everywhere, and hence $I\hat{h}_n = Ih_n$. The function $\hat{f}(x, y)$ coincides with $f(x, y)$ almost everywhere, and hence can be written in the form

$$\hat{f}(x, y) = f(x, y) + z(x, y),$$

where $z(x, y)$ is nonzero only on a set of measure zero. But by Steps 2 and 3, both $\hat{f}$ and z belong to Φ. Therefore, by Step 1, f also belongs to Φ, as asserted.

Step 5. Φ contains every function $\varphi \in L(W)$, since every such φ is the difference between two functions of the class $L^+(W)$. The proof of Fubini's theorem is now complete.

Remark 1. It is natural to ask whether the converse of Fubini's theorem holds, i.e., does the existence of the iterated integral

$$I_Y\{I_X\varphi(x, y)\} \tag{14}$$

imply that $\varphi(x, y)$ is summable on the set W? In general, $\varphi(x, y)$ will not be summable on W (see Probs. 7 and 8, p. 57). However, *if $\varphi(x, y)$ is measurable*[9] *and nonnegative, then the existence of* (14) *does in fact imply the summability of $\varphi(x, y)$ on W, together with the relation*

$$I\varphi = I_Y\{I_X\varphi(x, y)\}. \tag{15}$$

To see this, suppose the iterated integral

$$I_Y\{I_X\varphi(x, y)\} = A$$

exists, where

$$\varphi(x, y) = \lim_{n\to\infty} h_n(x, y).$$

Then the function

$$\varphi_n(x, y) = \min \{\varphi(x, y), \max (h_1, \ldots, h_n)\}$$

is also measurable, since

$$\varphi_n(x, y) = \lim_{m\to\infty} \min \{h_m, \max (h_1, \ldots, h_n)\},$$

and moreover $\varphi_n(x, y)$ is bounded by the summable function $\max (h_1, \ldots, h_n)$. Therefore $\varphi_n(x, y)$ is summable on W, by the result of Sec. 2.8.1, and

$$I\varphi_n = I_Y\{I_X\varphi_n(x, y)\} \leqslant A,$$

by Fubini's theorem. Since $\varphi_n \nearrow \varphi$ and $I\varphi_n \leqslant A$, it follows from Corollary 1 to Levi's theorem that the function φ is summable. But then, applying Fubini's theorem again, we obtain (15), as required.

---

[9] I.e., if $\varphi(x, y)$ is the almost-everywhere limit of a sequence of elementary functions, as in Sec. 2.8.1.

Remark 2. Here we have started from a "ready-made" integral on the Cartesian product W of the sets X and Y, related in a certain way to the integrals defined on X and Y themselves. In Sec. 6.8, we shall see how to *construct* the integral on W, starting from known integrals on X and Y satisfying the necessary constraints.

2.11. Integrals of Variable Sign

So far we have required that the elementary integral be nonnegative, and this fact has played a key role in our considerations. However, in analysis one also encounters the case where the continuous linear functional Ih,[10] which we wish to call the integral, can take values of either sign (e.g., the Stieltjes integral, the subject of a special study in Part 2).

2.11.1. Riesz's representation theorem. The structure of continuous linear functionals of variable sign is revealed by the following

THEOREM 6 (*Riesz's representation theorem*). *Every continuous linear functional Ih of variable sign,*[11] *defined on the space of elementary functions $h \in H$, can be represented as the difference between two nonnegative functionals on H.*

Proof. Denoting the set of all nonnegative $h \in H$ by H^+ (note that H^+ is not a linear space), we define the functional

$$Jh = \sup_{0 \leqslant k(x) \leqslant h(x)} Ik \tag{16}$$

on H^+. If the integral I were nonnegative as before, then obviously we would have $Jh = Ih$, but this can no longer be asserted. In any event, it is obvious that $Jh \geqslant 0$ and $Jh \geqslant Ih$, since we can always choose $k = 0$ or $k = h$. The possibility that $Jh = 0$ for all $h \in H^+$ is not excluded, nor do we exclude *a priori* the case where Jh takes the value $+\infty$ for certain $h \in H^+$ (however, see Step 2 below). The nub of the proof is to show that the (nonnegative) functional J is linear and continuous on H^+, and can be extended to the larger space H. This will be established in five steps.

Step 1. J is subadditive on H^+, i.e.,

$$J(h_1 + h_2) \leqslant Jh_1 + Jh_2 \tag{17}$$

for every $h_1, h_2 \in H^+$. In fact, suppose $0 \leqslant k_0 \leqslant h_1 + h_2$. Then k_0 can always be represented in the form $k_1 + k_2$, where $k_1 \leqslant h_1$, $k_2 \leqslant h_2$, since we need only set $k_1 = \min(h_1, k_0)$, $k_2 = k_0 - k_1$. But $k_0 = k_1 + k_2$

[10] Here linearity and continuity are defined as in Axioms 1 and 3, p. 24.

[11] Which are, of course, themselves linear and continuous.

implies $Ik_0 = Ik_1 + Ik_2 \leqslant Jh_1 + Jh_2$, and then (17) follows at once, after taking the least upper bound of the left-hand side.

Step 2. J is finite on H^+. Suppose, to the contrary, that $Jh_0 = +\infty$ for some $h_0 \in H^+$. Then, as we now show, there exists a sequence of functions $h_n \in H^+$ such that

$$h_n \leqslant \tfrac{1}{2}h_{n-1}, \qquad Jh_n = +\infty, \qquad |Ih_n| \geqslant n, \tag{18}$$

which contradicts the assumed continuity of the functional I, since $h_n(x) \searrow 0$. To start the induction, let h_0 be the first function. Assuming that functions $h_0, h_1, \ldots, h_{n-1}$ satisfying the conditions (18) have already been constructed, we choose a function k such that

$$0 \leqslant k \leqslant h_{n-1}, \qquad Ik \geqslant |Ih_{n-1}| + 2n$$

(here we use the fact that $Jh_{n-1} = \infty$). If $Jk = \infty$, we set $h_n = \frac{1}{2}k$, while if $Jk < \infty$, then certainly

$$J(h_{n-1} - k) = \infty, \qquad |I(k - h_{n-1})| \geqslant Ik - |Ih_{n-1}| \geqslant 2n,$$

which allows us to set $h_n = \frac{1}{2}(h_{n-1} - k)$. Thus a sequence satisfying (18) actually exists, and hence the assumption that $Jh_0 = +\infty$ leads to a contradiction, i.e., Jh is finite for all $h \in H^+$, as asserted.

Step 3. J is additive and positive homogeneous on H^+. First we prove that J is additive on H^+, i.e., that

$$J(h_1 + h_2) = Jh_1 + Jh_2 \tag{19}$$

for every $h_1, h_2 \in H^+$. Given any $\varepsilon > 0$, we choose functions $k_1 \leqslant h_1$ and $k_2 \leqslant h_2$ such that

$$Ik_1 \geqslant Jh_1 - \varepsilon, \qquad Ik_2 \geqslant Jh_2 - \varepsilon.$$

Then

$$J(h_1 + h_2) \geqslant I(k_1 + k_2) = Ik_1 + Ik_2 \geqslant Jh_1 + Jh_2 - 2\varepsilon,$$

and hence

$$Jh_1 + Jh_2 \leqslant J(h_1 + h_2), \tag{20}$$

since $\varepsilon > 0$ is arbitrary. But together (20) and (17) imply (19). Moreover, it is obvious from the definition (16) that J is positive homogeneous on H^+, i.e., that

$$J(\alpha h) = \alpha Jh \qquad (h \in H^+, \alpha > 0).$$

Step 4. J can be extended onto H. If

$$\varphi = h_1 - h_2,$$

where $\varphi \in H$ and $h_1, h_2 \in H^+$, we define

$$J\varphi = Jh_1 - Jh_2.$$

This definition is unique, since if $\varphi = h_1 - h_2 = h_3 - h_4$, where h_3, h_4 also belong to H^+, then

$$h_1 + h_4 = h_2 + h_3,$$
$$Jh_1 + Jh_4 = J(h_1 + h_4) = J(h_2 + h_3) = Jh_2 + Jh_3,$$

and hence

$$Jh_1 - Jh_2 = Jh_3 - Jh_4.$$

The functional J is still additive on H, since

$$\varphi_1 = h_1 - k_1, \qquad \varphi_2 = h_2 - k_2 \qquad (h_1, k_1, h_2, k_2 \in H^+)$$

implies

$$\varphi_1 + \varphi_2 = (h_1 + h_2) - (k_1 + k_2),$$
$$J(\varphi_1 + \varphi_2) = J(h_1 + h_2) - J(k_1 + k_2)$$
$$= Jh_1 + Jh_2 - Jk_1 - Jk_2 = J\varphi_1 + J\varphi_2.$$

Moreover, for any real α, we have $J(\alpha\varphi) = \alpha J(\varphi)$. This is obvious for $\alpha \geqslant 0$, and hence we need only consider the case $\alpha = -1$. But $\varphi = h - k$ $(h, k \in H^+)$ implies $-\varphi = k - h$, so that indeed

$$J(-\varphi) = J(k) - J(h) = -[J(h) - J(k)] = -J\varphi.$$

The fact that J is linear on H now follows at once.

Step 5. J is continuous on H, i.e.,

$$h_n \searrow 0 \quad \text{implies} \quad Jh_n \to 0.$$

Given any $\varepsilon > 0$, we choose numbers $\varepsilon_n > 0$ such that

$$\sum_{n=1}^{\infty} \varepsilon_n \leqslant \varepsilon$$

and functions k_n, $0 \leqslant k_n \leqslant h_n$, such that

$$Ik_n \geqslant Jh_n - \varepsilon_n.$$

Moreover, let

$$\bar{k}_n = \min(k_1, \ldots, k_n).$$

Then it is claimed that

$$Jh_n \leqslant I\bar{k}_n + \sum_{i=1}^{n} \varepsilon_i. \tag{21}$$

For $n = 1$, this follows from the definition of the function $\bar{k}_1 = k_1$. Suppose (21) holds for the values $i = 1, \ldots, n$. Obviously

$$\bar{k}_{n+1} = \min(\bar{k}_n, k_{n+1}),$$
$$\max(\bar{k}_n, k_{n+1}) + \min(\bar{k}_n, k_{n+1}) = \bar{k}_n + k_{n+1},$$

and hence

$$I(\max(\bar{k}_n, k_{n+1})) + I\bar{k}_{n+1} = I\bar{k}_n + Ik_{n+1} \geqslant I\bar{k}_n + Jh_{n+1} - \varepsilon_{n+1}. \tag{22}$$

On the other hand, observing that

$$\bar{k}_n \leqslant k_n \leqslant h_n, \qquad k_{n+1} \leqslant h_{n+1} \leqslant h_n,$$

and assuming that (21) holds, we have

$$I(\max(\bar{k}_n, k_{n+1})) \leqslant J(h_n) \leqslant I\bar{k}_n + \sum_{i=1}^{n} \varepsilon_i. \tag{23}$$

It follows from (22) and (23) that

$$I\bar{k}_n + Jh_{n+1} - \varepsilon_{n+1} - I\bar{k}_{n+1} \leqslant I\bar{k}_n + \sum_{i=1}^{n} \varepsilon_i,$$

and hence

$$Jh_{n+1} \leqslant I\bar{k}_{n+1} + \sum_{i=1}^{n+1} \varepsilon_i,$$

which completes the induction. At the same time, we obtain the relation

$$\overline{\lim_{n\to\infty}}\, Jh_n \leqslant \overline{\lim_{n\to\infty}}\, I\bar{k}_n + \varepsilon = \varepsilon,$$

since $\bar{k}_n \leqslant h_n \searrow 0$ and hence $I\bar{k}_n \to 0$. But then $Jh_n \to 0$, as asserted, since $\varepsilon > 0$ is arbitrary.

The rest of the proof is now straightforward. Consider the functional

$$N = J - I,$$

defined on H. If $h \geqslant 0$, then $Nh = Jh - Ih \geqslant 0$, so that N, like J, is a nonnegative functional. Since J and I are linear and continuous, so is N,[12] and the theorem is proved.

2.11.2. Construction of a space of summable functions for the functional *I*. Next we use the functionals J and N figuring in Theorem 6 to extend the domain of definition of the functional I. Actually, we start from a single nonnegative linear functional $K = J + N$, and follow the procedure described in Sec. 2.5. First we distinguish sets of K-measure zero. As on p. 24, a set $Z \subset X$ is said to be of K-measure zero if, given any $\varepsilon > 0$, there exists a nondecreasing sequence of nonnegative functions $h_n(x) \in H$ such that $Kh_n < \varepsilon$ and $\sup h_n(x) \geqslant 1$ on Z; every set of K-measure zero is automatically a set of J-measure zero and of N-measure zero, since $0 \leqslant Jh_n \leqslant Kh_n$, $0 \leqslant Nh_n \leqslant Kh_n$. Then we define the class L_K^+ consisting of functions $f(x)$ which are the limits of nondecreasing sequences $h_n \in H$ such that $Kh_n \leqslant C$ $(n = 1, 2, \ldots)$. Obviously, for such functions it makes sense to talk about Jf and Nf, in addition to

$$Kf = \lim_{n\to\infty} Kh_n.$$

---

[12] In particular, $h_n \searrow 0$ implies $Nh_n = Jh_n - Ih_n \to 0$ (actually $Nh_n \searrow 0$, since N is nonnegative).

Finally we form the class L_K from differences

$$\varphi = f - g \quad (f, g \in L_K^+).$$

If φ is such a function, the expressions $K\varphi = Kf - Kg$, $J\varphi$ and $N\varphi$ are all meaningful, and we can define the integral

$$I\varphi = J\varphi - N\varphi,$$

thereby extending the continuous linear functional I to the space L_K, where it satisfies the inequality

$$|I\varphi| \leqslant |J\varphi| + |N\varphi| = J\varphi^+ + J\varphi^- + N\varphi^+ + N\varphi^- = K\varphi^+ + K\varphi^- = K(|\varphi|).$$

2.11.3. Other representations of *I*. The canonical representation. In Theorem 6, we found a representation of the functional I of variable sign as a difference

$$I = J - N \tag{24}$$

between two nonnegative functionals. This representation is not unique. In fact, let L be any nonnegative continuous linear functional on H. Then, besides the representation (24), we can also write

$$I = (J + L) - (N + L). \tag{25}$$

It turns out that (25) is actually the most general representation of I as a difference between two nonnegative functionals. To see this, suppose we have any representation

$$I = J_1 - N_1, \tag{26}$$

where J_1 and N_1 are nonnegative continuous linear functionals. Then, given any nonnegative functions $h, k \in H$ such that $0 \leqslant k(x) \leqslant h(x)$,

$$Ik = J_1k - N_1k \leqslant J_1k \leqslant J_1h,$$

and hence

$$Jh = \sup_{0 \leqslant k(x) \leqslant h(x)} Ik \leqslant J_1h.$$

Therefore $J_1 = J + L$, where $L = J_1 - J$ is a nonnegative continuous linear functional. Moreover

$$N_1 = J_1 - I = J + L - I = N + L,$$

i.e., we have reduced (26) to the form (25), as required.

At the same time, we find that the representation (24), explicitly constructed in Theorem 6, has a simple characterization in the class of all possible representations of the form (26), i.e., the functionals J and N figuring in (24) are the *smallest possible* among all that can figure in (26). For this reason, (24) will be called the *canonical representation* of the functional I.

Finally, we show that the space L_K, $K = J + N$ corresponding to the canonical representation is the largest possible, in the sense that every function $\varphi \in L_{K_1}$, $K_1 = J_1 + N_1$ also belongs to L_K. Clearly, we need only verify that every function $f \in L^+_{K_1}$ also belongs to L^+_K. In fact, after adding a suitable elementary function (if necessary), f is the limit, except on a set of K_1-measure zero, of a sequence of nonnegative elementary functions $h_n(x)$ with bounded integrals $I_{K_1}h_n$. But obviously $I_K h_n \leqslant I_{K_1}h_n$ for nonnegative elementary functions, and hence the integrals $I_K h_n$ are also bounded. Moreover, the set of K_1-measure zero on which the sequence h_n fails to converge to f is also a set of K-measure zero, since $h \geqslant 0$, $I_{K_1}h < \varepsilon$ implies $I_K h < \varepsilon$. Therefore the sequence h_n converges to f everywhere except on a set of K-measure zero, i.e., $f \in L^+_K$, as required.

3

THE LEBESGUE INTEGRAL IN n-SPACE

In this chapter we shall use the general scheme of Chap. 2 to construct the Lebesgue integral for a finite-dimensional space, choosing as the elementary functions first step functions and then continuous functions.

3.1. Relation between the Riemann Integral and the Lebesgue Integral

It will be recalled from Sec. 1.6 that if $f(x)$ is Riemann integrable (over the basic block $\mathbf{B}$), then $f(x)$ is the almost-everywhere limit of a nondecreasing sequence of (lower) step functions. Suppose we choose the family H of elementary functions to be the family of step functions $h(x)$, with the "natural" definition of the integral, i.e.,

$$Ih = \sum_{j=1}^{m} h_j s(B_j), \qquad B_j = \{x\colon h(x) = h_j\}. \tag{1}$$

Then, as shown in Chap. 1, I satisfies all the axioms for an elementary integral given in Sec. 2.1. Therefore the entire scheme of Chap. 2 is applicable to the present case, and implies the existence of a linear space $L(\mathbf{B})$ of functions summable on the block $\mathbf{B}$. Moreover, there is a Lebesgue integral $I\varphi$ defined on $L(\mathbf{B})$, and $L(\mathbf{B})$ is complete when equipped with the norm $\|\varphi\| = I(|\varphi|)$.

We now try to form some idea (albeit partial) of the size of the class $L(\mathbf{B})$. It follows from the considerations of Chap. 1 that every Riemann-integrable function f (in particular, every continuous function) is also

Lebesgue-integrable (in fact, $f \in L^+$), and moreover the Lebesgue integral of f coincides with the Riemann integral of f. Thus the process of Lebesgue integration applies to every Riemann-integrable function. But it also applies to a much larger class of functions. For example, a function with no continuity points at all can still be Lebesgue integrable. Thus the Dirichlet function $\chi(x)$, defined in footnote 2, p. 8, although not Riemann integrable, is Lebesgue integrable (being nonzero only on a set of measure zero), and in fact $I\chi = 0$. There are more complicated examples of Lebesgue-integrable functions which have no continuity points even after an arbitrary alteration on a set of measure zero (see Prob. 4, p. 148).

It follows from Sec. 2.8.1 that *every bounded measurable function*[1] *is summable*, since in the present case, constants are summable functions. This immediately raises the question of whether there are bounded nonsummable functions. The answer is in the affirmative (see Prob. 6, p. 57), although no explicit example of such a function has yet been constructed.

3.2. Improper Riemann Integrals and the Lebesgue Integral

Next we consider functions which have improper Riemann integrals. First suppose $\varphi(x)$ is nonnegative and bounded in the block $\mathbf{B} = \{x \colon |x_j| \leqslant a_j, j = 1, \ldots, n\}$, everywhere except at the origin of coordinates, where $\varphi(x)$ becomes infinite, and suppose the (ordinary) Riemann integral

$$\int_{\mathbf{B}-B_\varepsilon} \varphi(x)\, dx \tag{2}$$

exists for every block of the form $B_\varepsilon = \{x \colon |x_j| \leqslant \varepsilon, j = 1, \ldots, n\}$, where (2) is defined in the obvious way by partitioning $\mathbf{B} - B_\varepsilon$ ($B_\varepsilon \subset \mathbf{B}$) into subblocks with no interior points in common. Then $\varphi(x)$ is said to be Riemann integrable on $\mathbf{B}$ if the integral (2) approaches a limit as $\varepsilon \to 0$, and this limit, denoted by

$$\int_{\mathbf{B}} \varphi(x)\, dx,$$

is called the (improper) Riemann integral of $\varphi(x)$. For example, the function

$$\left(\sum_{j=1}^{n} x_j^2\right)^{-\alpha/2}$$

has an improper Riemann integral over any block containing the origin if $\alpha < n$, but not if $\alpha \geqslant n$. Let us analyze this situation from the standpoint of the Lebesgue integral. The integral (2) is the integral over the whole block $\mathbf{B}$ of the function

$$\varphi_\varepsilon(x) = \begin{cases} \varphi(x) & \text{for } x \in \mathbf{B} - B_\varepsilon, \\ 0 & \text{for } x \in B_\varepsilon, \end{cases}$$

[1] I.e., every bounded function which is the limit almost everywhere of a sequence of step functions (see p. 37).

whose Riemann and Lebesgue integrals coincide. As $\varepsilon \to 0$, the functions $\varphi_\varepsilon(x)$ form a nondecreasing sequence converging to $\varphi(x)$. Therefore, if the integral (2) approaches a limit as $\varepsilon \to 0$, then, by Levi's theorem, the function φ is summable, with Lebesgue integral equal to the limit of (2) as $\varepsilon \to 0$, i.e., to the improper Riemann integral of φ. Conversely, if (2) approaches infinity as $\varepsilon \to 0$, i.e., if φ has no improper Riemann integral, then φ cannot have a Lebesgue integral $I\varphi$, since the existence of $I\varphi$ would imply $I\varphi_\varepsilon \leqslant I\varphi$ for all ε. In other words, φ is summable on $\mathbf{B}$ if and only if φ has an improper Riemann integral on $\mathbf{B}$.

The case of an unbounded domain of integration (rather than an unbounded integrand) is handled in much the same way. For example, let R_n denote all of Euclidean n-space, and, given a nonnegative function $\varphi(x)$, suppose the (ordinary) Riemann integral

$$\int_{B_r} \varphi(x)\,dx \tag{3}$$

exists for every block of the form $B_r = \{x: |x_j| \leqslant r, j = 1, \ldots, n\}$. Then $\varphi(x)$ is said to be Riemann integrable on R_n if the integral (3) approaches a limit as $r \to \infty$, and this limit, denoted by

$$\int_{R_n} \varphi(x)\,dx,$$

is called the (improper) Riemann integral of $\varphi(x)$. In terms of Lebesgue integrals, (3) is the integral over R_n of the function

$$\varphi_r(x) = \begin{cases} \varphi(x) & \text{for} \quad x \in B_r, \\ 0 & \text{for} \quad x \in R_n - B_r, \end{cases}$$

whose Riemann and Lebesgue integrals coincide. Here, to construct the Lebesgue integral, we choose as our elementary functions all step functions vanishing outside finite unions of (bounded) blocks. Thus φ is summable on R_n if and only if φ has an improper Riemann integral on R_n. The argument is the same as for the case where φ has a singular point, except that now B_r and φ_r play the roles of $\mathbf{B} - B_\varepsilon$ and φ_ε.

To recapitulate, *the class of summable functions contains all nonnegative functions φ with improper Riemann integrals.* It is essential that φ be nonnegative, since otherwise this assertion breaks down (see Probs. 4 and 5, p. 54).

3.3. Fubini's Theorem for Functions of Several Real Variables

We now examine the meaning of Fubini's theorem when applied to functions of several real variables. In the notation of Sec. 2.10, let X be the block

$$\mathbf{B}_X = \{x: a_1 \leqslant x_1 \leqslant b_1, \ldots, a_m \leqslant x_m \leqslant b_m\}$$

in m-space, and let Y be the block

$$\mathbf{B}_Y = \{y: c_1 \leqslant y_1 \leqslant d_1, \ldots, c_n \leqslant y_n \leqslant d_n\}$$

in n-space. Then $W = X \times Y$ is the block

$$\mathbf{B} = \{(x, y): a_j \leqslant x_j \leqslant b_j, c_k \leqslant y_k \leqslant d_k, j = 1, \ldots, m; k = 1, \ldots, n\}$$

in $(m + n)$-space. For the space $H(W)$ we choose all step functions $h(x, y)$ defined on the block $\mathbf{B}$, i.e., all functions of the form

$$h(x, y) = \sum_{j=1}^{m} \alpha_j \chi_{B_X^{(j)}}(x) \chi_{B_Y^{(j)}}(y),$$

where the blocks $B_X^{(j)}$ and $B_Y^{(j)}$ form partitions of $\mathbf{B}_X$ and $\mathbf{B}_Y$, respectively. The function $\chi_{B_X^{(j)}}(x)$ is the characteristic function of $B_X^{(j)}$, i.e.,

$$\chi_{B_X^{(j)}}(x) = \begin{cases} 1 & \text{for } x \in B_X^{(j)}, \\ 0 & \text{for } x \in \mathbf{B}_X - B_X^{(j)}, \end{cases}$$

and similarly for $\chi_{B_Y^{(j)}}(y)$. For the elementary integral on $H(W)$, we make the "natural choice"

$$Ih = \sum_{j=1}^{m} \alpha_j s(B_X^{(j)}) s(B_Y^{(j)}).$$

The space $H(W)$, equipped with this integral, clearly satisfies all the hypotheses of Fubini's theorem, i.e., every function $h(x, y) \in H(W)$ is summable in x for almost all y [being a step function in x except for finitely many sheets of discontinuity of the $\chi_{B_Y^{(j)}}(y)$], the integral

$$I_X h = \sum_{j=1}^{m} \alpha_j s(B_X^{(j)}) \chi_{B_Y^{(j)}}(y)$$

is summable in y (being a step function in y), and

$$Ih = \sum_{j=1}^{m} \alpha_j s(B_X^{(j)}) s(B_Y^{(j)}) = I_Y\{I_X h(x, y)\}.$$

It follows that the space $L(W)$ generated by $H(W)$ and I has the same three properties. In particular,

$$I\varphi = I_Y\{I_X \varphi(x, y)\}$$

for every $\varphi(x, y) \in L(W)$. Moreover, we can also write

$$I\varphi = I_X\{I_Y \varphi(x, y)\},$$

because of the symmetry between the roles of x and y in the definition of the elementary integral.

3.4. Continuous Functions as Elementary Functions, with the Riemann Integral as Elementary Integral

We now describe another way of constructing the space L of Lebesgue-integrable functions. Suppose we choose as our elementary functions the set $\tilde{H}$ of all continuous functions $f(x)$ in the (closed bounded) block $\mathbf{B}$, with the Riemann integral as elementary integral. Then $\tilde{H}$ satisfies Axioms a, b and the proposed elementary (Riemann) integral, henceforth denoted by $\tilde{I}f$, satisfies Axioms 1–3 (see p. 24). The only nontrivial part of this assertion is the verification of Axiom 3. But Axiom 3 is an immediate consequence of the estimate

$$|\tilde{I}f_m| = \left|\int_{\mathbf{B}} f_m(x)\,dx\right| \leqslant s(\mathbf{B}) \max_{x\in\mathbf{B}} |f_m(x)|$$

and the following

LEMMA (*Dini's lemma*). *A nonincreasing sequence of nonnegative continuous functions $f_m(x)$ converging to zero at every point of a closed bounded block* $\mathbf{B}$ *converges to zero uniformly in* $\mathbf{B}$.

Proof. Given any $\varepsilon > 0$ and any point $x_0 \in \mathbf{B}$, we can find an integer $m = m(x_0)$ such that $f_m(x_0) < \varepsilon$. Then we find a neighborhood $U(x_0)$ such that $f_m(x) < \varepsilon$ for all $x \in U(x_0)$. Obviously, if $p > m$, then $f_p(x) \leqslant f_m(x) < \varepsilon$ for all $x \in U(x_0)$. Constructing such a neighborhood for every point of $\mathbf{B}$, we obtain a covering of $\mathbf{B}$, from which we can select a *finite* subcovering (cf. p. 13). Let q be the smallest subscript of the functions participating in this subcovering. Then $f_r(x) < \varepsilon$ for every $x \in \mathbf{B}$ provided that $r > q$, and the lemma is proved.

Thus all the prerequisites for constructing a theory of the integral, based on continuous functions as elementary functions, with the Riemann integral $\tilde{I}$ as elementary integral, are satisfied. Let $\tilde{L}$ denote the corresponding space of summable functions, equipped with a Lebesgue integral $\tilde{I}f$. Then, as we now show, this construction of Lebesgue-integrable functions agrees with that of Sec. 3.1, based on step functions as elementary functions (with the obvious definition of elementary integral), leading to a space L of summable functions equipped with a Lebesgue integral If. More precisely, we prove the following

THEOREM. *The two constructions of the Lebesgue integral in n-space are equivalent, i.e., $\tilde{L} = L$ and $\tilde{I}f = If$.*

Proof. The proof will be established in four steps:

Step 1. Every continuous function $f(x)$ belongs to L. Given any $\varepsilon > 0$, we can find a partition $\Pi = \{B_1, \ldots, B_m\}$ of the basic block $\mathbf{B}$ so fine that

$$\left|\int_{\mathbf{B}} f(x)\,dx - \sum_{j=1}^{m} f(\xi_k)s(B_j)\right| < \varepsilon \qquad (\xi_j \in B_j),$$

or equivalently,

$$\left| \int_{\mathbf{B}} f(x)\, dx - Ih_\Pi(x) \right| < \varepsilon, \tag{4}$$

where $h_\Pi(x)$ is the step function equal to $f(\xi_j)$ in the block B_j. Since $h_\Pi(x)$ converges uniformly to $f(x)$ as the partition Π is refined indefinitely, i.e., as $d(\Pi) \to 0$, it follows from Lebesgue's theorem (see p. 36) that $f \in L$ and

$$If = \lim_{d(\Pi)\to 0} Ih_\Pi = \tilde{I}f,$$

where the last equality is implied by (4).

Step 2. Every step function $h(x)$ belongs to $\tilde{L}$. Every function equal to 1 in a block B and to 0 outside B, and hence every step function h, can be represented (in various ways) as the limit of a bounded everywhere convergent sequence of continuous functions $f_m(x)$,[2] where, as just shown, $\tilde{I}f_m = If_m$. Therefore, again by Lebesgue's theorem, $h \in \tilde{L}$ and

$$\tilde{I}h = \lim_{m\to\infty} \tilde{I}f_m = \lim_{m\to\infty} If_m = Ih.$$

Step 3. Both constructions lead to the same sets of measure zero. Let $\tilde{Z}$ be a set of measure zero relative to the integral $\tilde{I}$. Then, given any $\varepsilon > 0$, there exists a nondecreasing sequence of nonnegative continuous functions $f_m^{(\varepsilon)}(x)$ such that $\tilde{I}f_m^{(\varepsilon)} < \varepsilon$ and $\sup f_m^{(\varepsilon)}(x) \geqslant 1$ on $\tilde{Z}$. By Step 1, every $f_m^{(\varepsilon)} \in L$ and $If_m^{(\varepsilon)} = \tilde{I}f_m^{(\varepsilon)}$. Therefore, by Corollary 3 to Levi's theorem, $\tilde{Z}$ is a set of measure zero relative to I. Conversely, let Z be a set of measure zero relative to the integral I. Then, given any $\varepsilon > 0$, there exists a nondecreasing sequence of nonnegative step functions $h_m^{(\varepsilon)}(x)$ such that $Ih_m^{(\varepsilon)} < \varepsilon$ and $\sup h_m^{(\varepsilon)}(x) \geqslant 1$ on Z. By Step 2, every $h_m^{(\varepsilon)} \in \tilde{L}$ and $\tilde{I}h_m^{(\varepsilon)} = Ih_m^{(\varepsilon)}$. Therefore, by the same corollary, Z is a set of measure zero relative to $\tilde{I}$. Thus we have shown that the phrase "almost everywhere" means the same thing in the two spaces L and $\tilde{L}$.

Step 4. Monotonic passages to the limit and formation of differences. Suppose $f \in L^+$, so that f is the limit (almost everywhere) of a nondecreasing sequence of step functions h_m, with bounded integrals Ih_m. Then the integrals $\tilde{I}h_m = Ih_m$ are bounded, and hence, by Corollary 1 to Levi's theorem, $f \in \tilde{L}$ and $\tilde{I}f = If$. Conversely, suppose $f \in \tilde{L}^+$, so that f is the limit (almost everywhere) of a nondecreasing sequence of continuous functions f_m, with bounded integrals $\tilde{I}f_m$. Then the integrals $If_m = \tilde{I}f_m$ are bounded, and hence, by the same corollary, $f \in L$ and $If = \tilde{I}f$. Finally, taking differences, we find that $\tilde{L}$ contains every function $f \in L$, and *vice versa*, with $\tilde{I}f = If$. This completes the proof.

[2] Details on the construction of such functions $f_m(x)$ will be given on p. 98.

PROBLEMS

1. Suppose $f(x)$ equals 1 on an open set $G \subset [a, b]$ and 0 on the complement of G. Show that $f(x)$ belongs to $L^+[a, b]$.

Hint. If $G = \bigcup_{j=1}^{\infty} \Delta_j$, then $f(x) = \sum_{j=1}^{\infty} h_j(x)$, where $h_j(x) = 1$ on Δ_j and 0 outside Δ_j.

2. Construct an open set $G \subset [a, b]$ such that the function $f(x)$ equal to 0 on G and 1 on its complement does not belong to $L^+[a, b]$.

Hint. Choose

$$G = \bigcup_{j=1}^{\infty} \Delta_j = \bigcup_{j=1}^{\infty} (\alpha_j, \beta_j), \qquad \sum_{j=1}^{\infty} (\beta_j - \alpha_j) < b - a,$$

where G is such that every point $x \in [a, b]$ is a limit point of G.

3. Suppose a summable function $f(x)$, defined on the closed interval $[a, b]$, vanishes outside $[\alpha, \beta]$, where $a < \alpha < \beta < b$, so that $f(x)$ can be "shifted." Prove that $f(x)$ is "continuous in the mean," in the sense that given any $\varepsilon > 0$, there exists a $\delta > 0$ such that

$$\|f(x + \Delta x) - f(x)\| < \varepsilon \quad \text{if} \quad |\Delta x| < \delta, \tag{5}$$

where $\| \ \|$ denotes the L-norm.

Hint. Show that the set of all f satisfying (5) is closed in L. Then verify (5) for step functions.

4. Consider the function $f(x) = x^\alpha \sin (x^\beta)$, defined on the half-open interval $(0, 1]$. For what values of the real parameters α and β is $f(x)$

a) Riemann-integrable (in the improper sense);
b) Lebesgue-integrable?

Ans. a) For $\alpha > -1 - |\beta|$; b) For $\alpha > -1 - \beta$ ($\beta > 0$) and $\alpha > -1$ ($\beta < 0$).

Comment. For $\beta < 0$, $\beta - 1 < \alpha < -1$, the function $f(x)$ has an improper Riemann integral, but no Lebesgue integral.

5. Let $f(x)$ be the same as in the preceding problem, but this time defined on the infinite interval $(1, \infty)$. For what values of α and β is $f(x)$

a) Riemann integrable (in the improper sense);
b) Lebesgue integrable?

Ans. a) For $\alpha < |\beta| - 1$; b) For $\alpha < -1$ ($\beta > 0$) and $\alpha < -1 - \beta$ ($\beta < 0$).

Comment. For $\beta > 0$, $-1 < \alpha < -1 + \beta$, the function $f(x)$ has an improper Riemann integral, but no Lebesgue integral.

6 (*A nonmeasurable function*). To simplify the construction, imagine the interval [0, 1] wrapped around a circle Γ of circumference 1, and measure all distances along the circle. Two points ξ and η of the circle Γ will be called "like" if the distance between them is rational, and "unlike" if the distance is irrational. The countable set of all points like a given point (i.e., at rational distances from the point) will be called a "class." The set of all points of the circle is the union of an uncountable collection of disjoint classes. Let $f(x)$ be a function defined on Γ which for every class takes the value 1 for one member of the class and the value 0 for all other members. Show that $f(x)$ cannot be measurable.

Hint. If $f(x)$ is measurable, then it is summable, and so are all its "translates" $f(x + h)$, with $If(x + h) = If(x)$. Show that

$$\sum_r f(x + r) \equiv 1,$$

where r ranges over all rational numbers. By Levi's theorem,

$$\sum_r f(x + r) = \sum_r If(x) = I1 = 1,$$

which is incompatible with either $If > 0$ or $If = 0$.

7. Consider the double integrals

a) $\displaystyle\int_0^\infty\int_0^\infty e^{-xy}\sin x \sin y\, dx\, dy$; b) $\displaystyle\int_0^1\int_0^1 \frac{x^2 - y^2}{(x^2 + y^2)^2}\, dx\, dy.$

Show that the corresponding iterated integrals exist, for either order of integration, and are the same in Case a but different in Case b. Show that nevertheless the double integrals do not exist, i.e., that the integrands are not summable. Why doesn't this contradict Fubini's theorem (more precisely, its converse, as stated in Remark 1, p. 43)?

Hint. The integrands do not have constant sign.

8.[3] We start from the following two postulates of set theory:

1. The points of the continuum $C = \{0 \leqslant \xi \leqslant 1\}$ can be ordered by a new relation $\prec$ such that every subset has a least element (the *well-ordering hypothesis*).
2. With this ordering, every subset $\{\xi : \xi \prec \xi_0\}$, for arbitrary $\xi_0 \in C$, has no more than countably many elements (the *continuum hypothesis*).

Consider the set E of all points (x, y) of the square $0 \leqslant x \leqslant 1$, $0 \leqslant y \leqslant 1$ satisfying the "inequality" $x \prec y$. Show that every horizontal cross section of E (i.e., every intersection of E with a horizontal straight line) contains no more than countably many points, while the same is true of the complement of every vertical cross section. Show that the characteristic function $h(x, y)$ of the set E (i.e., the function equal to 1 on E and 0 otherwise) satisfies the relations

$$I_Y\{I_X h(x, y)\} = 0, \qquad I_X\{I_Y h(x, y)\} = 1.$$

Why doesn't this contradict Remark 1, p. 43 (the converse of Fubini's theorem)?

Hint. The function $h(x, y)$ is, of course, nonmeasurable.

---

[3] Due to G. P. Tolstov.

Part 2

THE STIELTJES INTEGRAL

4

THE RIEMANN-STIELTJES INTEGRAL

4.1. Blocks and Sheets

In this chapter, we introduce the *Riemann-Stieltjes integral*, which generalizes the ordinary Riemann integral in n-space and allows us to take account of "spatial inhomogeneity" (in a sense that will soon be apparent). As a first step, we modify the concept of a block, defined in Sec. 1.1 as a set of points $x = (x_1, \ldots, x_n)$ satisfying the inequalities

$$a_1 \leqslant x_1 \leqslant b_1, \ldots, a_n \leqslant x_n \leqslant b_n. \tag{1}$$

In Part 1 only the size and volume of a block mattered, so that it made no difference which of the symbols $\leqslant$ or $<$ appeared in (1). However, we now intend to replace the volume by the more general notion of a "Stieltjes quasi-volume," which can "concentrate" on the boundary of a block or even at individual points. (In fact, we shall even allow quasi-volumes to take negative values.) Therefore, from now on, we must be more careful about just what is meant by a block.

As before, the basic block **B** is a set of points of the form (1), but now we permit **B** to be infinite (we were moving in this direction in Sec. 3.2), i.e., one or more of the numbers $a_1, b_1, \ldots, a_n, b_n$ are allowed to be infinite ($-\infty$ for the a_j, $+\infty$ for the b_j), and correspondingly, **B** may contain points at infinity. Thus the basic block is always compact in the "natural topology," and hence any sequence of points in **B** contains a subsequence converging to a finite or infinite limit.[1]

---

[1] In this regard, we observe that an infinite block can always be transformed into a finite block by the substitution $x_j = \tan \xi_j$ $(j = 1, \ldots, n)$, where one or more end points of the new block $\mathbf{B}^*$, consisting of points $\xi = (\xi_1, \ldots, \xi_n)$, is equal to $\pm\pi/2$ (for further details, see Sec. 4.4.3).

By a *sheet* we mean the intersection of **B** with some hyperplane of dimension $n - 1$ parallel to a coordinate hyperplane (i.e., with some surface $x_j = \text{const}$). Note that some or all of the points of a sheet may be points at infinity. Given a basic block

$$\mathbf{B} = \{x: a_1 \leqslant x_1 \leqslant b_1, \ldots, a_n \leqslant x_n \leqslant b_n\}, \tag{2}$$

the set

$$\Gamma_k = \{x: a_1 \leqslant x_1 \leqslant b_1, \ldots, x_k = a_k, \ldots, a_n \leqslant x_n \leqslant b_n$$

is called the *k'th sheet of the lower boundary* of **B**, and the union of all the Γ_k $(k = 1, \ldots, n)$ is called the (*complete*) *lower boundary* of **B**. Similarly, the set

$$\Gamma^{(k)} = \{x: a_1 \leqslant x_1 \leqslant b_1, \ldots, x_k = b_k, \ldots, a_n \leqslant x_n \leqslant b_n\}$$

is called the *k'th sheet of the upper boundary* of **B**, and the union of all the $\Gamma^{(k)}$ $(k = 1, \ldots, n)$ is called the (*complete*) *upper boundary* of **B**. The set of all points at infinity belonging to either the lower or the upper boundary of **B** will be called the *improper boundary* of **B**.

Next let $\alpha_1, \beta_1, \ldots, \alpha_n, \beta_n$ be any numbers satisfying the inequalities

$$a_j \leqslant \alpha_j < \beta_j \leqslant b_j \qquad (j = 1, \ldots, n),$$

where a_j and b_j are the same as in (2). Then by a *subblock* of the basic block **B** we mean a set of the form

$$B = \{x: \alpha_1 < x_1 \leqslant \beta_1, \ldots, \alpha_n < x_n \leqslant \beta_n\} \tag{3}$$

if $\alpha_j > a_j$ for every j, but if $\alpha_j = a_j$ for any j, we replace the inequality $\alpha_j < x_j$ by $\alpha_j \leqslant x_j$. In other words, B contains any point of the lower boundary of **B** which is a limit point of B. (Note that the definition allows the basic block to be a subblock of itself!) The word "block," without further qualification, can mean either the basic block **B** or a subblock of **B**. The empty set $\varnothing$ will also be regarded as a block.

Remark. It should be noted that the definition of a subblock $B \subset \mathbf{B}$ depends on whether or not the intersection of the closure of B with the lower boundary of **B** is empty (in fact, the intersection is always adjoined to B). This could be avoided by defining all blocks, including the basic block itself, as "half-open" sets of the form (3). Admittedly, this choice achieves a certain notational simplicity at this point, but it would become unmanageable later, primarily because the use of Dini's lemma (p. 54) depends on **B** being compact (see Prob. 6, p. 108).

The following two properties of blocks are easily verified:

1) If B and B' are blocks, then so is their intersection BB';
2) If B_1 and B are blocks $B_1 \subset B$, then there exist blocks $B_2, \ldots, B_m$ such that

$$B = B_1 \cup B_2 \cup \cdots \cup B_m,$$

where the blocks $B_1, B_2, \ldots, B_m$ are (pairwise) disjoint.

Remark. In the language of Sec. 7.1, the set of all blocks (contained in a basic block **B**) forms a *semiring.* Note that Property 2 fails if blocks are defined as in Part 1, with $<$ replaced by $\leqslant$ in (3).

Finally, we introduce the notion of a *dense set of blocks.* Given a basic block (2), in each closed interval $a_k \leqslant x_k \leqslant b_k$, we choose a dense subset E_k $(k = 1, \ldots, n)$, which in particular contains the end points a_k and b_k. Then the set Q of all blocks of the form (3) with end points $\alpha_1, \beta_1, \ldots, \alpha_n, \beta_n$ belonging to the sets $E_1, \ldots, E_n$, respectively, is said to be *dense* in **B**. In this way, every collection of dense sets $E_1, \ldots, E_n$ generates a dense set of blocks $B \subset \mathbf{B}$.

4.2. Quasi-Volumes

Given a dense set of blocks Q in the basic block **B** (which may be infinite), suppose a real number $\sigma(B)$ is associated with every block $B \in Q$, where $\sigma(B)$ is *additive* in the sense that

$$\sigma(B) = \sigma(B_1) + \cdots + \sigma(B_m)$$

if B is a union of disjoint blocks $B_1, \ldots, B_m$ [in particular, $\sigma(\varnothing) = 0$]. Then the function $\sigma(B)$ is called a (*Stieltjes*) *quasi-volume.* It is important to note that in general $\sigma(B)$ is *signed*, i.e., can take values of either sign.

If $\sigma(B) \geqslant 0$ for every block $B \in Q$, the quasi-volume is said to be *nonnegative.* A quasi-volume σ is said to be of *bounded variation* if, given any partition Π of the block **B** into a set of disjoint blocks $B_1, \ldots, B_m \in Q$, the inequality

$$\sum_{j=1}^{m} |\sigma(B_j)| \leqslant C \tag{4}$$

holds, where the constant C does not depend on the choice of the blocks $B_1, \ldots, B_m$. The smallest value of the constant C figuring in (4), i.e., the quantity

$$V_{\mathbf{B}}(\sigma) = \sup_{\Pi} \sum_{j=1}^{m} |\sigma(B_j)|,$$

where the least upper bound is taken with respect to *all* partitions of the block **B**, is called the *total variation of the quasi-volume* σ *in the block* **B**. If the quasi-volume σ is nonnegative, i.e., if $\sigma(B) \geqslant 0$ for every $B \in Q$, the condition (4) reduces to finiteness of $\sigma(\mathbf{B})$.

We now give some examples of quasi-volumes, defined for all subblocks $B \subset \mathbf{B}$:

Example 1. Let

$$\sigma(B) = s(B),$$

where $s(B)$ is the volume of the block B in the ordinary sense (and **B** is finite).

Example 2. Let

$$\sigma(B) = \int_B g(x)\,dx,$$

where $g(x)$ is a function summable over **B**. If $g(x) \geqslant 0$, then $\sigma(B)$ is a nonnegative quasi-volume. In the general case, where $g(x)$ has variable sign, $\sigma(B)$ is of bounded variation, since

$$\sum_{j=1}^{m} |\sigma(B_j)| = \sum_{j=1}^{m} \left| \int_{B_j} g(x)\,dx \right| \leqslant \int_{\mathbf{B}} |g(x)|\,dx$$

if $B = B_1 \cup \cdots \cup B_m$. (Supply some missing details.)

Example 3. Given a sequence of points $c_1, \ldots, c_m, \ldots$ in the basic block **B** and a sequence of real numbers $g_1, \ldots, g_m, \ldots$ such that

$$\sum_{m=1}^{\infty} |g_m| = g < \infty,$$

let $\sigma(B)$ equal the sum of all the g_m such that the corresponding points c_m belong to the block B. If all the $g_m \geqslant 0$, then $\sigma(B)$ is nonnegative. In the general case, where the numbers g_m take either sign, $\sigma(B)$ has total variation no greater than g.

4.3. Quasi-Length and the Generating Function

In the case $n = 1$, the block **B** reduces to a closed interval $[a, b]$, where one or both end points can be infinite. A subblock $B \subset \mathbf{B}$ is then a half-open interval $(\alpha, \beta]$ if $\alpha > a$ or a closed interval if $\alpha = a$. Thus, in writing $(\alpha, \beta]$, we do so with the understanding that (is to be replaced by [if $\alpha = a$. To specify a dense set of blocks Q, we choose a set of points E which is dense in $[a, b]$ and contains the end points a and b. Then the block $B = (\alpha, \beta]$ belongs to Q if and only if both points α and β belong to E. In the case of intervals, it is more natural to refer to quasi-volume as *quasi-length*. With every quasi-length $\sigma(\alpha, \beta]$, $\alpha, \beta \in E$, we can associate a function of the real variable x, defined by

$$F(x) = \begin{cases} \sigma[a, x] & \text{for } x \in E, x \neq a, \\ 0 & \text{for } x = a. \end{cases}$$

Then, from a knowledge of the function $F(x)$, we can find the quasi-length of every $(\alpha, \beta] \in Q$, i.e.,

$$\sigma(\alpha, \beta] = \sigma[a, \beta] - \sigma[a, \alpha] = F(\beta) - F(\alpha). \tag{5}$$

The function $F(x)$ will be called the *generating function* of the quasi-length σ (synonymously, the *distribution function* of σ).

Conversely, let $F(x)$ be any function which is finite on E and vanishes for $x = a$. Then $F(x)$ can serve as a generating function, since the interval function $\sigma(\alpha, \beta]$ *defined* in terms of $F(x)$ by using formula (5) is automatically additive.

Obviously, the quasi-length $\sigma(\alpha, \beta]$ is nonnegative if and only if the corresponding generating function $F(x)$ is nondecreasing on E. We now interpret the definition of a quasi-length of bounded variation in terms of the generating function. In the present case, a partition of the block $\mathbf{B} = [a, b]$ into disjoint subblocks $B_1, \ldots, B_m$ corresponds to a partition of the closed interval $[a, b]$ into half-open subintervals (except for the first), i.e.,

$$[a, b] = [x_0, x_1] \cup (x_1, x_2] \cup \cdots \cup (x_{m-1}, x_m],$$

where $a = x_0 < x_1 < \cdots < x_{m-1} < x_m = b$ and the points $x_0, x_1, \ldots, x_{m-1}, x_m$ all belong to E. Then the inequality (4) on p. 63 takes the form

$$\sum_{j=1}^{m} |\sigma(B_j)| = \sum_{j=1}^{m} |F(x_j) - F(x_{j-1})| \leqslant C, \tag{6}$$

where $F(x_0) = 0$. A function $F(x)$ defined on the set E satisfying the inequality (6) for any choice of the integer m and the points $x_0, \ldots, x_m$ in E will be called a *function of bounded variation.* Thus a quasi-length σ is of bounded variation if and only if its generating function $F(x)$ is of bounded variation. The smallest value of the constant C figuring in (6), i.e., the quantity

$$V_a^b(F) = \sup \sum_{j=1}^{m} |F(x_j) - F(x_{j-1})|,$$

where the least upper bound is taken with respect to all partitions of the interval $[a, b]$, is called the *total variation* of the generating function F in the interval $[a, b]$.

Remark. The generating function can also be defined for an n-dimensional quasi-volume, although in this case it is not a particularly useful concept. For simplicity, we consider the case $n = 2$. Then the basic block $\mathbf{B}$ is a set defined by inequalities of the form

$$a_1 \leqslant x_1 \leqslant b_1, \qquad a_2 \leqslant x_2 \leqslant b_2.$$

Suppose E_1 is dense in $[a_1, b_1]$ and contains the points a_1, b_1, and similarly for E_2. Let $B_{\alpha_1\alpha_2}^{\beta_1\beta_2}$ be the block defined by

$$\alpha_1 < x_1 \leqslant \beta_1, \qquad \alpha_2 < x_2 \leqslant \beta_2, \qquad \alpha_1, \beta_1 \in E_1, \qquad \alpha_2, \beta_2 \in E_2,$$

where $\alpha_j < x_j$ is replaced by $\alpha_j \leqslant x_j$ if $\alpha_j = a_j$ $(j = 1, 2)$. If we define the function

$$F(x_1, x_2) = \sigma(B_{a_1a_2}^{x_1x_2}), \tag{7}$$

then the formula

$$\begin{aligned}\sigma(B_{\alpha_1\alpha_2}^{\beta_1\beta_2}) &= \sigma(B_{a_1a_2}^{\beta_1\beta_2}) - \sigma(\beta_{a_1a_2}^{\alpha_1\beta_2}) - \sigma(B_{a_1a_2}^{\beta_1\alpha_2}) + \sigma(\beta_{a_1a_2}^{\alpha_1\alpha_2}) \\ &= F(\beta_1, \beta_2) - F(\alpha_1, \beta_2) - F(\beta_1, \alpha_2) + F(\alpha_1, \alpha_2)\end{aligned}$$

allows us to reconstruct the quasi-volume $\sigma(B)$ of any block $B \in Q$. For this reason, the function (7) is called the *generating function* of the quasi-volume $\sigma(B)$.

4.4. The Riemann-Stieltjes Integral and Its Properties

We now introduce a far-reaching generalization of the concept of the Riemann integral, studied in Chap. 1.

4.4.1. Construction of the Riemann-Stieltjes integral. First we assume that the basic block $\mathbf{B}$ is finite and that a quasi-volume $\sigma(B)$ of bounded variation is defined on some dense set Q of blocks $B \subset \mathbf{B}$. Let $f(x)$ be a function defined on the block $\mathbf{B}$. Consider an arbitrary partition Π of the block $\mathbf{B}$ into disjoint subblocks belonging to Q, i.e., $\mathbf{B} = B_1 \cup \cdots \cup B_m$, and as in Sec. 1.1, let $d(\Pi)$ denote the largest size of the blocks $B_1, \ldots, B_m$. Choosing a point ξ_j in each block B_j, we form the *Riemann-Stieltjes sum*

$$S_\Pi(f) = \sum_{j=1}^{m} f(\xi_j)\sigma(B_j). \tag{8}$$

Let $\Pi_1, \ldots, \Pi_p, \ldots$ be a sequence of partitions such that $d(\Pi_p) \to 0$, and suppose the sequence $S_{\Pi_p}(f)$ has a limit as $p \to \infty$, which is independent of the choice of the sequence Π_p [provided only that $d(\Pi_p) \to 0$] or of the points $\xi_j \in B_j$. Then the limit is called the *Riemann-Stieltjes integral* of the function $f(x)$ [over the block $\mathbf{B}$] *with respect to the quasi-volume* σ, and we write

$$I_\sigma f \equiv \int_{\mathbf{B}} f(x)\sigma\,(dx) = \lim_{d(\Pi)\to 0} S_\Pi(f).$$

Correspondingly, the function $f(x)$ is said to be *Riemann-Stieltjes integrable* (over the block $\mathbf{B}$) *with respect to the quasi-volume* σ.

THEOREM 1. *If $f(x)$ is continuous in the block $\mathbf{B}$, then $f(x)$ is Riemann-Stieltjes integrable over $\mathbf{B}$ with respect to any quasi-volume σ of bounded variation.*

Proof. By hypothesis, given any $\varepsilon > 0$, we can find a partition $\Pi = \{B_j\}$ of the basic block so fine that $|f(x') - f(x'')| < \varepsilon$ for all x', $x'' \in B_j$. Such a partition will be said to *belong to* ε. Given a partition belonging to ε and a finer partition $\tilde{\Pi} = \{B_{jk}\}$, where $B_{jk} \subset B_j$ for every k, let ξ_{jk} be an arbitrary point of the block B_{jk}. Then clearly

$$f(\xi_{jk}) - f(\xi_j) = \varepsilon_{jk}$$

has absolute value less than ε. By definition,

$$S_\Pi(f) = \sum_j f(\xi_j)\sigma(B_j), \qquad S_{\tilde{\Pi}}(f) = \sum_{j,k} f(\xi_{jk})\sigma(B_{jk}),$$

which implies

$$|S_\Pi(f) - S_{\tilde{\Pi}}(f)| = \left| \sum_j \left[\sum_k f(\xi_{jk})\sigma(B_{jk}) - f(\xi_j)\sigma(B_j) \right] \right|$$

$$= \left| \sum_j \left[\sum_k \varepsilon_{jk}\sigma(B_{jk}) \right| < \varepsilon \sum_{j,k} |\sigma(B_{jk})| \leqslant \varepsilon V_{\mathbf{B}}(\sigma).$$

It follows that if Π and Π' are any two partitions belonging to ε, and if Π^* is the new partition consisting of all intersections of blocks of Π with blocks of Π', then

$$|S_\Pi(f) - S_{\Pi^*}(f)| < \varepsilon V_{\mathbf{B}}(\sigma), \qquad |S_{\Pi'}(f) - S_{\Pi^*}(f)| < \varepsilon V_{\mathbf{B}}(\sigma),$$

and hence

$$|S_\Pi(f) - S_{\Pi'}(f)| < 2\varepsilon V_{\mathbf{B}}(\sigma). \tag{9}$$

Now let Π_p be a sequence of partitions belonging to a sequence of numbers $\varepsilon_p \to 0$. Then, applying the above argument to the partitions Π_p and Π_{p+q}, we see that the numbers $S_{\Pi_p}(f)$ form a Cauchy sequence, and hence tend to some limit $I_\sigma f$. If Π'_p is another sequence of partitions belonging to the numbers ε_p, then, applying the same argument, this time to the partitions Π_p and Π'_p, we find that

$$S_{\Pi_p}(f) - S_{\Pi'_p}(f) \to 0.$$

Thus $I_\sigma f$ is independent of the choice of the sequence Π_p, and the theorem is proved. We note, in passing, that the inequality

$$\left| \int_{\mathbf{B}} f(x)\sigma(dx) - \sum_{j=1}^{m} f(\xi_j)\sigma(B_j) \right| \leqslant \varepsilon V_{\mathbf{B}}(\sigma) \tag{10}$$

holds for any partition belonging to the number ε.

Remark. Let $\sigma(\alpha, \beta]$ be a quasi-length of bounded variation. Then Theorem 1 implies the existence of the Riemann-Stieltjes integral

$$I_\sigma f = \int_{[a,b]} f(x)\sigma(dx) = \lim \sum_{j=1}^{m} f(\xi_j)\sigma(x_{j-1}, x_j] \tag{11}$$

for every function $f(x)$ continuous in the interval $[a, b]$. Here, of course, ξ_j is an arbitrary point in $(x_{j-1}, x_j]$, and in the limit on the right, the maximum length of the subintervals $(x_{j-1}, x_j]$ is made to approach zero. Let $F(x)$ be the generating function of the quasi-length $\sigma(\alpha, \beta]$. Then another way of writing the integral (11) is

$$\int_{[a,b]} f(x)\, dF(x),$$

suggested by the relation

$$\lim \sum_{j=1}^{m} f(\xi_j)\sigma(x_{j-1}, x_j] = \lim \sum_{j=1}^{m} f(\xi_j)[F(x_j) - F(x_{j-1})]$$

[again with $F(x_0) = F(a) = 0$]. In particular, if $f(x) \equiv 1$, we have

$$\int_{[a,b]} 1 \cdot dF(x) = \sigma[a, b] = F(b) = F(b) - F(a),$$

in keeping with elementary calculus.

4.4.2. Further properties. The following two properties of the Riemann-Stieltjes integral obviously hold in the general case (i.e., without any special assumptions about the continuity of the integrand):

1) If $f_1(x)$ and $f_2(x)$ are Riemann-Stieltjes integrable, then so is $\alpha_1 f_1(x) + \alpha_2 f_2(x)$, where α_1 and α_2 are arbitrary real numbers, and moreover

$$\int_{\mathbf{B}} [\alpha_1 f_1(x) + \alpha_2 f_2(x)]\sigma(dx) = \alpha_1 \int_{\mathbf{B}} f_1(x)\sigma(dx) + \alpha_2 \int_{\mathbf{B}} f_2(x)\sigma(dx).$$

2) If $f(x)$ is Riemann-Stieltjes integrable and if $|f(x)| \leqslant M$ for all $x \in \mathbf{B}$, then

$$\left| \int_{\mathbf{B}} f(x)\sigma(dx) \right| \leqslant MV_{\mathbf{B}}(\sigma). \tag{12}$$

In some cases, the Riemann-Stieltjes integral can be expressed in terms of the ordinary Riemann integral. Thus consider Example 2, p. 64, where

$$\sigma(B) = \int_B g(x)\, dx,$$

and suppose the summable function $g(x)$ is itself continuous in $\mathbf{B}$. Then, given any continuous function $f(x)$, we have

$$\int_{\mathbf{B}} f(x)\sigma(dx) = \int_{\mathbf{B}} f(x)g(x)\, dx. \tag{13}$$

In fact, if $\Pi = \{B_j\}$ is a partition of $\mathbf{B}$ belonging to ε for *both* $f(x)$ and $f(x)g(x)$, then, according to (10),

$$\left| \int_{\mathbf{B}} f(x)\sigma(dx) - \sum_{j=1}^{m} f(\xi_j)\sigma(B_j) \right| \leqslant \varepsilon V_{\mathbf{B}}(\sigma),$$

$$\left| \int_{\mathbf{B}} f(x)g(x)\, dx - \sum_{j=1}^{m} f(\xi_j)g(\xi_j)s(B_j) \right| \leqslant \varepsilon s(\mathbf{B}),$$

where $s(B)$ is ordinary volume, and hence

$$\left| \sum_{j=1}^{m} f(\xi_j)\sigma(B_j) - \sum_{j=1}^{m} f(\xi_j)g(\xi_j)s(B_j) \right|$$

$$= \left| \sum_{j=1}^{m} f(\xi_j) \int_{B_j} [g(x) - g(\xi_j)]\, dx \right| \leqslant \varepsilon s(\mathbf{B}) \max_{x \in \mathbf{B}} |f(x)|.$$

But then

$$\left| \int_{\mathbf{B}} f(x)\sigma(dx) - \int_{\mathbf{B}} f(x)g(x)\,dx \right| \leqslant \varepsilon \left\{ V_{\mathbf{B}}(\sigma) + s(\mathbf{B}) + s(\mathbf{B}) \max_{x \in \mathbf{B}} |f(x)| \right\},$$

which implies (13), since ε is arbitrary.

4.4.3. The case of infinite B. The above construction can easily be carried over to the case where the block **B** is infinite. It is only necessary to be careful about the meaning of an "arbitrarily fine partition." Suppose we make the transformation $x_j = \tan \xi_j$ already mentioned in footnote 1, p. 61. Then **B** goes into a finite block $\mathbf{B}^*$ in the space of points $\xi = (\xi_1, \ldots, \xi_n)$, and we say that the partition Π of the block **B** is "arbitrarily fine" if the corresponding partition Π^* of the block $\mathbf{B}^*$ is arbitrarily fine in the usual sense, i.e., if $d(\Pi^*)$ can be made less than any preassigned $\varepsilon > 0$. Moreover, a function $f(x) = f(x_1, \ldots, x_n)$ defined in **B** is said to be *continuous in* **B** if the function $f(\tan \xi_1, \ldots, \tan \xi_n)$ is continuous in $\mathbf{B}^*$. In other words, if Γ is the improper boundary of **B**, then $f(x)$ is continuous in $\mathbf{B} - \Gamma$ and can be continuously extended onto Γ. Once these conventions have been established, we see at once that Theorem 1 remains valid for an infinite block **B**.

As we now show, Theorem 1 still holds even if $f(x)$ cannot be continuously extended onto the improper boundary of **B**, provided the conditions on σ are strengthened somewhat. First we need the following

DEFINITION. *Let $\sigma(B)$ be a quasi-volume of bounded variation defined on a dense set of blocks Q in an infinite block* **B**. *Then $\sigma(B)$ is said to be continuous at infinity, if, given any $\varepsilon > 0$, there exists a finite block $B_\varepsilon \subset \mathbf{B}$, $B_\varepsilon \in Q$ such that*

$$\sum_{j=1}^{p} |\sigma(B_j)| < \varepsilon \tag{14}$$

for arbitrary disjoint blocks B_j contained in $\mathbf{B} - B_\varepsilon$ $(B_j \in Q)$.[2]

Example. Let $\mathbf{B} = [0, \infty]$ and consider the quasi-volume $\sigma(\alpha, \beta]$ with generating function

$$F(x) = \begin{cases} e^{-x} & \text{for} \quad 0 \leqslant x < \infty,\ x \text{ rational}, \\ 1 & \text{for} \quad x = \infty. \end{cases}$$

Then σ is of bounded variation, but not continuous at infinity, since, for example, $\sigma(x, \infty] > \frac{1}{2}$ for arbitrarily large rational x. However, to make σ continuous at infinity, we need only "correct" $F(x)$ by setting $F(\infty) = 0$.

We are now in a position to prove the promised refinement of Theorem 1:

THEOREM 2. *Let* **B** *be an infinite block with improper boundary Γ, and suppose $f(x)$ is continuous and bounded in $\mathbf{B} - \Gamma$. Then $f(x)$ is Riemann-Stieltjes integrable over* **B** *with respect to any quasi-volume σ of bounded variation which is continuous at infinity.*

---

[2] It is important to note that the blocks B_j can be infinite.

Proof. The theorem has just been proved (without the extra condition on σ) for the case where $f(x)$ can be continuously extended onto Γ. Therefore the reader should now have in mind an example like $f(x) = \cos x\ (-\infty < x < \infty)$, $\mathbf{B} = [-\infty, \infty]$, where $f(x)$ cannot be continuously extended onto Γ. Given any $\varepsilon > 0$, let $B_\varepsilon \subset \mathbf{B}$, $B_\varepsilon \in Q$ be such that the inequality (14) holds. Observing that any partition of the basic block $\mathbf{B}$ generates a partition of B_ε (consisting of the intersections of B_ε with the blocks of the partition), let $\Pi = \{B_j\}$ and $\Pi' = \{B_j'\}$ be any two partitions of $\mathbf{B}$ such that both partitions $\{B_jB_\varepsilon\}$ and $\{B_j'B_\varepsilon\}$ belong to ε (in the sense defined on p. 66). Such partitions exist, since $f(x)$ is uniformly continuous in B_ε. Let ξ_j be any (finite) point of B_j' and ξ_j' any point of B_j'. Then, according to formula (9), p. 67,

$$\left|\sum_j f(\xi_j)\sigma(B_jB_\varepsilon) - \sum_j f(\xi_j')\sigma(B_j'B_\varepsilon)\right| < 2\varepsilon V_{\mathbf{B}}(\sigma), \tag{15}$$

where $V_{\mathbf{B}}(\sigma)$ is the total variation of σ. Moreover,

$$\sum_j f(\xi_j)\sigma(B_j) - \sum_j f(\xi_j)\sigma(B_jB_\varepsilon)$$
$$= \sum_j f(\xi_j)[\sigma(B_j) - \sigma(B_jB_\varepsilon)] = \sum_j f(\xi_j)\sigma\Big(\bigcup_k B_{jk}\Big),$$

where the B_{jk} are suitable disjoint blocks contained in $\mathbf{B} - B_\varepsilon$ ($B_{jk} \in Q$) whose union equals $B_j - B_jB_\varepsilon$ (if $B_j \subset B_\varepsilon$, all the B_{jk} are empty). It follows that

$$\left|\sum_j f(\xi_j)\sigma(B_j) - \sum_j f(\xi_j)\sigma(B_jB_\varepsilon)\right| \leqslant M\sum_{j,k}|\sigma(B_{jk})| \leqslant M\varepsilon, \tag{16}$$

where

$$M = \sup_{x\in\mathbf{B}-\Gamma} |f(x)|,$$

and similarly

$$\left|\sum_j f(\xi_j')\sigma(B_j') - \sum_j f(\xi_j')\sigma(B_j'B_\varepsilon)\right| \leqslant M\varepsilon. \tag{17}$$

Combining (15), (16) and (17), we obtain

$$\left|\sum_j f(\xi_j)\sigma(B_j) - \sum_j f(\xi_j')\sigma(B_j')\right| \leqslant 2\varepsilon[V_{\mathbf{B}}(\sigma) + M].$$

Since this effectively generalizes formula (9), p. 67 to the present case, the rest of the proof is identical with that of Theorem 1.

Remark. The class of functions which are integrable with respect to a given quasi-volume σ also contains discontinuous functions, and criteria for Riemann-Stieltjes integrability, analogous to those given in Chap. 1, could be established. However, we shall not bother to do so, since the Lebesgue-Stieltjes integral, to be constructed later (see Chap. 5), leads to a much larger class of integrable functions than the Riemann-Stieltjes integral.

4.4.4. Equivalent quasi-volumes: a preview. There are cases where *different* quasi-volumes σ_1 and σ_2, defined on different dense sets of blocks $Q_1 = Q(\sigma_1)$ and $Q_2 = Q(\sigma_2)$, or even on the same set $Q = Q_1 = Q_2$, lead to the same Riemann-Stieltjes integral, in the sense that

$$\int_{\mathbf{B}} f(x)\sigma_1(dx) = \int_{\mathbf{B}} f(x)\sigma_2(dx) \tag{18}$$

for every function $f(x)$ continuous in $\mathbf{B}$.

Example. Let $n = 1$ and suppose the quasi-length $\sigma_1(\alpha, \beta]$ equals zero for every half-open interval $(\alpha, \beta]$ which does not contain a given point c $(a < c < b)$ or else contains c as an interior point. Moreover, suppose $\sigma_1(\alpha, c] = +1$ for every $\alpha < c$, while $\sigma_1(c, \beta] = -1$ for every $\beta > c$. On the other hand, let $\sigma_2(\alpha, \beta]$ be the quasi-length identically equal to zero. Then it is easily verified that

$$\int_{[a,b]} f(x)\sigma_1(dx) = \int_{[a,b]} f(x)\sigma_2(dx) = 0$$

for every function $f(x)$ continuous in $[a, b]$.

Two quasi-volumes σ_1 and σ_2 satisfying the condition (18) are said to be *equivalent*. The subject of equivalent quasi-volumes will be studied in detail in Sec. 5.7. For the time being, we merely anticipate some results showing the reader what is at issue:

1) *Every quasi-volume σ of bounded variation is equivalent to a quasi-volume $\tilde{\sigma}$, also of bounded variation, which is defined on all blocks $B \subset \mathbf{B}$ and is upper continuous in the sense that*

$$\tilde{\sigma}(B) = \lim_{m\to\infty} \tilde{\sigma}(B_m)$$

for any sequence of blocks B_m converging downward to the block B (symbolically $B_m \searrow B$).[3]

2) *Two equivalent upper continuous quasi-volumes coincide on all blocks $B \subset \mathbf{B}$.*

3) *The quasi-volume $\tilde{\sigma}$ is determined from the quasi-volume σ by the formula*

$$\tilde{\sigma}(B) = \lim_{m\to\infty} \sigma(B_m),$$

where $B_m \in Q(\sigma)$ and $B_m \searrow B$.

[3] Roughly speaking, the boundaries of B_m move downwards, approaching those of B from above, in a sense to be made precise in Sec. 5.5.

4.5. Essential Convergence. The Helly Theorems

Given a sequence of quasi-volumes $\sigma_1, \ldots, \sigma_m, \ldots$ defined on the sub-blocks of a basic block **B**, we would like to define the concept of convergence $\sigma_m \to \sigma$ in such a way that the formula

$$\lim_{m\to\infty} \int_{\mathbf{B}} f(x)\sigma_m(dx) = \int_{\mathbf{B}} f(x)\sigma(dx) \tag{19}$$

holds for every function continuous in **B**. The appropriate definition turns out to be the following: Given a sequence of quasi-volumes $\sigma_1, \ldots, \sigma_m, \ldots$ and another quasi-volume σ, all defined on the same dense set of blocks Q in **B**, we say that σ_m is *essentially convergent* to σ if

1) The total variations $V_{\mathbf{B}}(\sigma_m)$ form a bounded sequence;
2) For every $B \in Q$,

$$\lim_{m\to\infty} \sigma_m(B) = \sigma(B).$$

Clearly, the quasi-volume $\sigma(B)$ is of bounded variation, like the quasi-volumes $\sigma_m(B)$ themselves. In fact, given any partition of the basic block **B** into disjoint blocks $B_1, \ldots, B_p \in Q$, we have

$$\sum_{j=1}^{p} |\sigma(B_j)| = \lim_{m\to\infty} \sum_{j=1}^{p} |\sigma_m(B_j)| \leqslant C,$$

and since this estimate does not depend on the choice of the partition, the assertion is proved.

THEOREM 3 (*Helly's convergence theorem*). *If the sequence of quasi-volumes* $\sigma_1, \ldots, \sigma_m, \ldots$ *is essentially convergent to the quasi-volume* σ, *then the limit relation* (19) *holds for every function* $f(x)$ *continuous in* **B**.

Proof. Given any $\varepsilon > 0$, let $\Pi = \{B_j\}$, $B_j \in Q$ be a partition of **B** so fine that Π belongs to ε in the sense defined on p. 66. Then, according to formula (10), we have

$$\left| \int_{\mathbf{B}} f(x)\sigma(dx) - \sum_j f(\xi_j)\sigma(B_j) \right| \leqslant \varepsilon V_{\mathbf{B}}(\sigma),$$

$$\left| \int_{\mathbf{B}} f(x)\sigma_m(dx) - \sum_j f(\xi_j)\sigma_m(B_j) \right| \leqslant \varepsilon V_{\mathbf{B}}(\sigma_m).$$

Moreover, let N be an integer so large that

$$\left| \sum_j f(\xi_j)\sigma(B_j) - \sum_j f(\xi_j)\sigma_m(B_j) \right| < \varepsilon$$

for all $m > N$. Combining the last three inequalities, we obtain

$$\left| \int_{\mathbf{B}} f(x)\sigma(dx) - \int_{\mathbf{B}} f(x)\sigma_m(dx) \right| < \varepsilon[V_{\mathbf{B}}(\sigma) + V_{\mathbf{B}}(\sigma_m) + 1] \leqslant (2C + 1)\varepsilon \tag{20}$$

for all $m > N$, where

$$C = \sup \{V_{\mathbf{B}}(\sigma), V_{\mathbf{B}}(\sigma_1), V_{\mathbf{B}}(\sigma_2), \ldots\}. \tag{21}$$

But this implies (19), since ε is arbitrary and C is independent of m.

Remark 1. Formulas (19) and (20) can be written more concisely as

$$\lim_{m \to \infty} I_{\sigma_m} f = I_\sigma f$$

and

$$|I_\sigma f - I_{\sigma_m} f| < (2C + 1)\varepsilon. \tag{22}$$

Remark 2. Theorem 3 can be generalized somewhat by allowing the function $f(x)$ to depend on m. In fact, if σ_m is a sequence of quasi-volumes converging essentially to a quasi-volume σ and if f_m is a sequence of continuous functions converging uniformly to a (continuous) function f, then

$$\lim_{m \to \infty} I_{\sigma_m} f_m = I_\sigma f.$$

This is an immediate consequence of (22) and the estimates

$$|I_{\sigma_m}(f - f_m)| \leqslant C \max_{x \in \mathbf{B}} |f - f_m|,$$

$$|I_\sigma(f - f_m)| \leqslant C \max_{x \in \mathbf{B}} |f - f_m|,$$

involving the same constant C defined by (21).

Remark 3. Suppose the block $\mathbf{B}$ is infinite, with improper boundary Γ. Then Theorem 3 remains valid if $f(x)$ is continuous and bounded in $\mathbf{B} - \Gamma$, even if $f(x)$ cannot be continuously extended onto Γ, provided the sequence σ_m is *equicontinuous at infinity* in the following sense (which is the natural generalization of the definition on p. 69): Given any $\varepsilon > 0$, there exists a finite block $B_\varepsilon \subset \mathbf{B}$, $B_\varepsilon \in Q$ such that

$$\sum_{j=1}^{p} |\sigma_m(B_j)| < \varepsilon \tag{23}$$

for arbitrary disjoint blocks B_j contained in $\mathbf{B} - B_\varepsilon$ $(B_j \in Q)$ *and all m*. In fact, let $B_\varepsilon \subset \mathbf{B}$, $B_\varepsilon \in Q$ be such that (23) holds, and let $\Pi = \{B_j\}$ be a partition of $\mathbf{B}$ belonging to the number ε. Then, according to formula (16), p. 70,

$$\left| \sum_j f(\xi_j)\sigma_m(B_j) - \sum_j f(\xi_j)\sigma_m(B_j B_\varepsilon) \right| \leqslant M\varepsilon,$$

where

$$M = \sup_{x \in \mathbf{B} - \Gamma} |f(x)|.$$

Moreover,

$$\left| \sum_j f(\xi_j)\sigma(B_j) - \sum_j f(\xi_j)\sigma(B_j B_\varepsilon) \right| \leqslant M\varepsilon,$$

since passage to the limit $m \to \infty$ in (23) gives

$$\sum_{j=1}^{p} |\sigma(B_j)| \leqslant \varepsilon.$$

It follows that

$$\begin{aligned} &\left| \sum_j f(\xi_j)\sigma(B_j) - \sum_j f(\xi_j)\sigma_m(B_j) \right| \\ &\qquad \leqslant 2M\varepsilon + \left| \sum_j f(\xi_j)\sigma(B_j B_\varepsilon) - \sum_j f(\xi_j)\sigma_m(B_j B_\varepsilon) \right| \\ &\qquad = 2M\varepsilon + \left| \sum_j f(\xi_j)[\sigma(B_j B_\varepsilon) - \sigma_m(B_j B_\varepsilon)] \right| \leqslant (2M + 1)\varepsilon, \end{aligned}$$

provided m is sufficiently large. Letting $d(\Pi) \to 0$, we obtain

$$|I_\sigma f - I_{\sigma_m} f| \leqslant (2M + 1)\varepsilon$$

for sufficiently large m, where the existence of the integrals is guaranteed by Theorem 2. Therefore

$$\lim_{m \to \infty} I_{\sigma_m} f = I_\sigma f,$$

as required.

THEOREM 4 (*Helly's selection principle*). *Let* $\Sigma = \{\sigma_\alpha(B)\}$ *be an infinite family of quasi-volumes, all defined on the same dense set of blocks* Q *in* $\mathbf{B}$, *where*

$$V_{\mathbf{B}}(\sigma_\alpha) \leqslant C$$

for every $\sigma_\alpha \in \Sigma$. *Then* Σ *contains a sequence* $\sigma_m(B)$ *which is essentially convergent to a quasi-volume* $\sigma(B)$ *of bounded variation.*

Proof. Clearly, Q contains a *sequence* of blocks $B_1, B_2, \ldots$ which is dense in $\mathbf{B}$. Since the set of numbers $\sigma_\alpha(B_1)$ is bounded, there is a sequence of quasi-volumes $\sigma_{1m} \in \Sigma$ such that the numerical sequence $\sigma_{1m}(B_1)$ is convergent. From the sequence σ_{1m} we can select a subsequence σ_{2m} such that $\sigma_{2m}(B_2)$, as well as $\sigma_{2m}(B_1)$, is convergent. Continuing this construction, given any integer p, we can find a sequence of quasi-volumes σ_{pm} such that the numerical sequences $\sigma_{pm}(B_1), \ldots, \sigma_{pm}(B_p)$ all converge (as $m \to \infty$). Therefore the diagonal sequence $\sigma_m \equiv \sigma_{mm}$ converges on all the blocks $B_1, B_2, \ldots$ The (essential) limit $\sigma(B)$ of the sequence $\sigma_m(B)$, defined on all the blocks $B_1, B_2, \ldots$, obviously represents a quasi-volume. Since $\sigma(B)$ is of bounded variation, by the argument given on p. 72, the proof is now complete.

Remark. For the case $n = 1$, the above results can all be paraphrased in terms of generating functions. Thus, given a sequence of generating functions $F_1, \ldots, F_m, \ldots$ and another generating function F, all defined on the same dense set $E \subset [a, b]$, we say that F_m is *essentially convergent* to F if

1) The total variations $V_a^b(F_m)$ form a bounded sequence;
2) For every $x \in E$,

$$\lim_{m \to \infty} F_m(x) = F(x).$$

Clearly. the limit function $F(x)$ is of bounded variation, like the generating functions $F_m(x)$ themselves. According to Helly's convergence theorem, *if the sequence of generating functions $F_m(x)$ is essentially convergent to the function $F(x)$, then*

$$\lim_{m \to \infty} \int_a^b f(x)\, dF_m(x) = \int_a^b f(x)\, dF(x) \tag{24}$$

for every function continuous in the interval $[a, b]$, while, according to Helly's selection principle, *if $\mathscr{F} = \{F_\alpha(x)\}$ is an infinite family of generating functions, all defined on the same dense set $E \subset [a, b]$, where $V_a^b(F_\alpha) \leqslant C$ for every $F_\alpha \in \mathscr{F}$, then $\mathscr{F}$ contains a sequence $F_m(x)$ which is essentially convergent to a generating function $F(x)$ of bounded variation.* Moreover, (24) continues to hold if the function $f(x)$ itself depends on m, or if the interval $[a, b]$ is infinite.

*4.6. Applications to Analysis

The theorems on passage to the limit inside a Stieltjes integral have numerous applications to mathematical analysis. We now digress to point some of these out.

4.6.1. Herglotz's theorem. Consider the problem of finding the general form of a function $w = f(z)$ which is analytic in the unit disk $|z| < 1$ and has a nonnegative real part, i.e., which maps the unit disk into the right half-plane. First we note that $w = \alpha + i\beta$ ($\alpha \geqslant 0$) is such a function, and so is

$$f_t(x) = \frac{e^{it} + z}{e^{it} - z},$$

where t is an arbitrary real number. In fact, for fixed t and $|z| < 1$, the points $z_1 = e^{it} + z$ and $z_2 = e^{it} - z$ lie inside the circle Γ of unit radius with center at the point e^{it}. Moreover, z_1 and z_2 lie on the same diameter of Γ, while the origin O lies on Γ itself (see Figure 1). But, by elementary geometry, the whole diameter subtends the angle $\pi/2$ at O, and hence the segment of the diameter joining the points z_1 and z_2 subtends an angle $\alpha < \pi/2$. In other words,

$$|\arg f_t(z)| = |\arg z_1 - \arg z_2| < \frac{\pi}{2},$$

which implies $\operatorname{Re} f_t(z) > 0$, as asserted.

Remarkably enough, it turns out that *every* function $w = f(z)$ analytic in the unit disk with a nonnegative real part is a "Stieltjes combination" of the particularly simple functions $f_t(z)$, $0 \leqslant t \leqslant 2\pi$:

THEOREM 5 (*Herglotz's theorem*). *If $f(z)$ is analytic in the disk $|z| < 1$ and has a nonnegative real part, then $f(z)$ can be represented in the form*

$$f(z) = \int_0^{2\pi} \frac{e^{it} + z}{e^{it} - z}\, dF(t) + i\beta, \tag{25}$$

where β is a real number and $F(t)$ is a bounded nondecreasing function.

Proof. As is well known, a function $f(z)$ analytic in the disk $|z| \leqslant r < 1$ can be represented in terms of the boundary values of its real part $u(z)$ by using Schwarz's formula[4]

$$f(z) = \frac{1}{2\pi} \int_0^{2\pi} \frac{re^{it} + z}{re^{it} - z}\, u(re^{it})\, dt + i\beta \qquad (|z| < r).$$

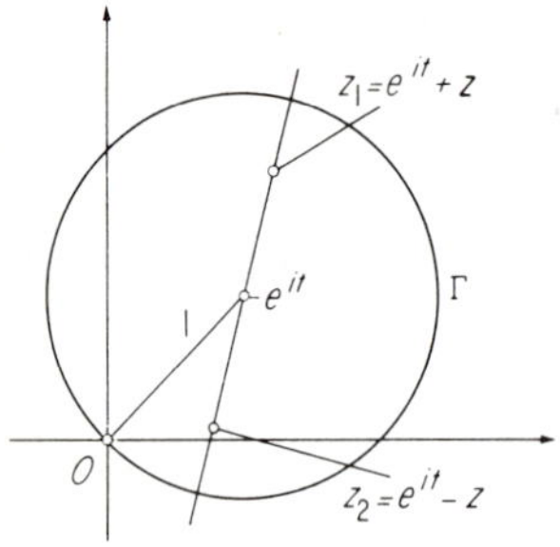

FIGURE 1

The integral on the right can be written in the form

$$f(z) = \int_0^{2\pi} \frac{re^{it} + z}{re^{it} - z}\, dF_r(t) + i\beta,$$

where

$$F_r(t) = \frac{1}{2\pi} \int_0^t u(re^{i\tau})\, d\tau$$

is a nondecreasing function of t. Moreover, by a familiar property of harmonic functions,[5] we have

$$F_r(t) \leqslant F_r(2\pi) = \frac{1}{2\pi} \int_0^{2\pi} u(re^{i\tau})\, d\tau = u(0).$$

---

[4] See e.g., A. I. Markushevich, *Theory of Functions of a Complex Variable, Volume II* (translated by R. A. Silverman), Prentice-Hall, Inc., Englewood Cliffs, N.J. (1965), Theorem 5.5, Corollary.

[5] I.e., $u(0)$ equals its average on the circle $|z| = r$ (see e.g., A. I. Markushevich, *op. cit.*, Theorem 5.6).

Therefore the total variations $V_0^{2\pi}(F_r)$, $0 < r < 1$ form a bounded set. Let $\{r_n\}$ be an increasing sequence of numbers in the interval (0, 1), exceeding $|z|$ and converging to 1. Then it follows from Helly's selection principle that the sequence of generating functions $\{F_{r_n}(t)\}$ contains a subsequence $\{F_{r_{n'}}(t)\}$ which is essentially convergent to a bounded nondecreasing generating function $F(t)$. Moreover, for fixed z, the sequence of functions

$$\frac{r_{n'}e^{it} + z}{r_{n'}e^{it} - z} \qquad (n = 1, 2, \ldots),$$

each continuous in the interval $0 \leqslant t \leqslant 2\pi$, converges uniformly in the same interval to the limit function

$$\frac{e^{it} + z}{e^{it} - z}.$$

Therefore, by Helly's convergence theorem

$$f(z) = \lim_{n' \to \infty} \int_0^{2\pi} \frac{r_{n'}e^{it} + z}{r_n e^{it} - z} \, dF_{r_{n'}}(t) + i\beta = \int_0^{2\pi} \frac{e^{it} + z}{e^{it} - z} \, dF(t) + i\beta$$

(cf. Remark 2, p. 73), and the theorem is proved. Note that the representation (25) includes the case $w = \alpha + i\beta$ ($\alpha \geqslant 0$) if we set

$$F(t) = \frac{\alpha t}{2\pi}.$$

4.6.2. Bernstein's theorem. A function $f(x)$ is said to be *completely monotonic* in the interval $[a, b]$ if all its derivatives exist and satisfy the inequalities

$$(-1)^n f^{(n)}(x) \geqslant 0 \qquad (n = 0, 1, 2, \ldots) \tag{26}$$

for every $x \in [a, b]$. A nonnegative constant is completely monotonic, and so is the function $e^{-\alpha x}$ ($\alpha > 0$). As we now show,[6] it turns out that *every* completely monotonic function in the interval $x \geqslant 0$ is a "Stieltjes combination" of the particularly simple functions $e^{-\alpha x}$ ($\alpha \geqslant 0$).

LEMMA 1. *If $f(x)$ satisfies* (26), *then*

$$\lim_{x \to \infty} x^n f^{(n)}(x) = 0, \tag{27}$$

and moreover

$$\frac{1}{n!} \int_0^\infty x^n |f^{(n+1)}(x)| \, dx = f(0) - f(\infty) < \infty \qquad (n = 0, 1, 2, \ldots). \tag{28}$$

[6] Following B. I. Korenblyum, *On two theorems from the theory of absolutely monotonic functions* (in Russian), Usp. Mat. Nauk, **6**, no. 4, 172 (1951).

Proof. According to (26), every function $(-1)^n f^{(n)}(x)$ is nonincreasing. Therefore, for $x > 0$,

$$|f^{(n)}(x)| = (-1)^n f^{(n)}(x) \leqslant \frac{2}{x}\int_{x/2}^{x}(-1)^n f^{(n)}(x)\,dx$$

$$= \frac{2}{x}\,|f^{(n-1)}(x) - f^{(n-1)}(x/2)|\,, \qquad (n = 1, 2, \ldots),$$

and (27) follows by induction, since clearly $f(x) - f(x/2) \to 0$ as $x \to \infty$. Moreover, integrating

$$(-1)^{n+1}\int_0^\infty x^n f^{(n+1)}(x)\,dx = \int_0^\infty x^n\,|f^{(n+1)}(x)|\,dx$$

by parts n times, and using (27), we obtain

$$\frac{1}{n!}\int_0^\infty x^n\,|f^{(n+1)}(x)|\,dx = \frac{(-1)^{n+1}}{n!}\int_0^\infty x^n f^{(n+1)}(x)\,dx$$

$$= -\int_0^\infty f'(x)\,dx = f(0) - f(\infty) < \infty,$$

which agrees with (28).

LEMMA 2. *If*

$$\varphi_n(x) = \begin{cases} \left(1 - \dfrac{x}{n}\right)^n & \text{for } 0 \leqslant x < n, \\ 0 & \text{for } n \leqslant x \leqslant \infty, \end{cases}$$

then

$$\lim_{n\to\infty} \varphi_n(x) = e^{-x}$$

uniformly for $x \geqslant 0$.

Proof. For $n > 1$, the function $|\varphi_n(x) - e^{-x}|$ achieves its maximum at the point a_n $(0 < a_n < \infty)$, where

$$\varphi_n'(a_n) + e^{-a_n} = 0,$$

i.e., where

$$\left(1 - \frac{a_n}{n}\right)^{n-1} = e^{-a_n} \qquad (0 < a_n < n).$$

Therefore

$$|\varphi_n(x) - e^{-x}| \leqslant |\varphi_n(a_n) - e^{-a_n}| = \frac{1}{n}\,a_n e^{-a_n} \leqslant \frac{1}{ne} \qquad (n = 2, 3, \ldots)$$

for all $x \in [0, \infty]$, and the lemma is proved.

THEOREM 6 (*Bernstein's theorem*). *If $f(x)$ is completely monotonic in the interval $x \geqslant 0$, then $f(x)$ can be represented in the form*

$$f(x) = \int_0^\infty e^{-\alpha x}\,dF(\alpha) + C, \tag{29}$$

where C is a nonnegative constant and $F(\alpha)$ is a bounded nondecreasing function.

Proof. Using (27), integrating by parts n times and making certain obvious changes of variables, we obtain

$$\begin{aligned} f(x) - f(\infty) &= -\int_x^\infty f'(t)\,dt = \frac{(-1)^{n+1}}{n!}\int_x^\infty (t-x)^n f^{(n+1)}(t)\,dt \\ &= \frac{(-1)^{n+1}n}{n!}\int_{x/n}^\infty \left(1 - \frac{x}{nt}\right)^n (nt)^n f^{(n+1)}(nt)\,dt \\ &= \int_0^\infty \varphi_n\left(\frac{x}{t}\right) d\left\{\frac{1}{n!}\int_0^t (-1)^{n+1} n(nt)^n f^{(n+1)}(nt)\,dt\right\} \\ &= \int_0^\infty \varphi_n(\alpha x)\,dF_n(\alpha) \qquad (n = 1, 2, \ldots), \end{aligned}$$

where φ_n is the same as in Lemma 2, and

$$F_n(\alpha) = \frac{1}{n!}\int_{1/\alpha}^\infty (-1)^{n+1} n(nt)^n f^{(n+1)}(nt)\,dt$$

[the integrals are all absolutely convergent because of (28)]. Clearly, every $F_n(\alpha)$ is bounded and nondecreasing, and moreover the total variations $V_0^\infty(F_n)$ form a bounded sequence, since

$$\begin{aligned} V_0^\infty(F_n) &= \frac{1}{n!}\int_0^\infty n(nt)^n \,|f^{(n+1)}(nt)|\,dt \\ &= \frac{1}{n!}\int_0^\infty t^n \,|f^{(n+1)}(t)|\,dt = f(0) - f(\infty) \qquad (n = 1, 2, \ldots). \end{aligned}$$

Therefore, by Helly's selection principle, the sequence of generating functions $F_n(\alpha)$ contains a subsequence $F_{n'}(\alpha)$ which is essentially convergent to a bounded nondecreasing generating function $F(\alpha)$. It follows from Helly's convergence theorem that[7]

$$f(x) - f(\infty) = \lim_{n'\to\infty}\int_0^\infty \varphi_{n'}(\alpha x)\,dF_{n'}(\alpha) = \int_0^\infty e^{-\alpha x}\,dF(\alpha),$$

i.e.,

$$f(x) = \int_0^\infty e^{-\alpha x}\,dF(\alpha) + f(\infty),$$

which agrees with (29), since obviously $f(\infty) \geqslant 0$. The term $f(\infty)$ can be incorporated into the integral, if we replace $F(\alpha)$ by the generating function

$$G(\alpha) = \begin{cases} f(\infty) + F(\alpha) & \text{for } \alpha > 0, \\ 0 & \text{for } \alpha = 0. \end{cases}$$

---

[7] Recall Remark 2, p. 73, observing that the functions $\varphi_n(\alpha x)$ and $e^{-\alpha x}$ are continuous at ∞ (for fixed x).

A closely related concept is that of an absolutely monotonic function, i.e., a function $f(x)$ is said to be *absolutely monotonic* in the interval $[a, b]$ if all its derivatives exist and are nonnegative for every $x \in [a, b]$:

$$f^{(n)}(x) \geqslant 0 \qquad (n = 0, 1, 2, \ldots).$$

The corresponding version of Bernstein's theorem is then

THEOREM 6'. *If $f(x)$ is absolutely monotonic in the interval $x \leqslant 0$, then $f(x)$ can be represented in the form*

$$f(x) = \int_0^\infty e^{\alpha x}\, dF(\alpha) + C,$$

where C is a nonnegative constant and $F(\alpha)$ is a bounded nondecreasing function.

Proof. If $f(x)$ is absolutely monotonic in the new sense, then $f(-x)$ is completely monotonic in the sense of the definition (26).

4.6.3. The Bochner-Khinchin theorem. A complex-valued function $f(x)$ is said to be *positive definite* in the interval $[a, b]$ if, given any n real numbers $x_1, \ldots, x_n$ in $[a, b]$ (where n itself is arbitrary), the $n \times n$ matrix $\|f(x_j - x_k)\|$ is positive definite, i.e., the quadratic form

$$\sum_{j,k=1}^{n} f(x_j - x_k)\xi_j\bar{\xi}_k$$

is nonnegative for arbitrary complex numbers $\xi_1, \ldots, \xi_n$ (the overbar denotes the complex conjugate). Any nonnegative constant is positive definite in any interval, and so is the function $e^{i\alpha x}$ (α real), since

$$\sum_{j,k=1}^{n} e^{i\alpha(x_j - x_k)}\xi_j\bar{\xi}_k = \sum_{j=1}^{n} e^{i\alpha x_j}\xi_j \overline{\sum_{k=1}^{n} e^{i\alpha x_k}\xi_k} = \left|\sum_{j=1}^{n} e^{i\alpha x_j}\xi_j\right|^2 \geqslant 0.$$

It turns out that *every* positive definite function defined on the real line is a "Stieltjes combination" of the particularly simple positive definite functions $e^{i\alpha x}$. In fact, the celebrated *Bochner-Khinchin theorem* asserts that *if $f(x)$ is positive definite and continuous for all x, then $f(x)$ can be represented in the form*

$$f(x) = \int_{-\infty}^{\infty} e^{i\alpha x}\, dF(\alpha),$$

where $F(\alpha)$ is a bounded nondecreasing function. Since, for fixed x, the function $e^{i\alpha x}$ cannot be continuously extended onto the improper boundary of the interval $[-\infty, \infty]$, the proof requires extra care, and, as might be expected, considerations like those given in Remark 3, p. 73 play a role. We omit the details, which would lead us too far afield.[8]

---

[8] See G. E. Shilov, *Mathematical Analysis, A Special Course* (translated and edited by J. D. Davis and D. A. R. Wallace), Pergamon Press, Inc., New York (1965), p. 438.

4.7. Structure of Signed Quasi-Volumes

We now study the relation between *signed* quasi-volumes (i.e., quasi-volumes which can take values of either sign) and nonnegative quasi-volumes. The perceptive reader will note the analogy between this section and Sec. 2.11.

4.7.1. Representation of a signed quasi-volume σ as the difference between two nonnegative quasi-volumes. As already noted in Sec. 4.2, a nonnegative quasi-volume $p(B)$ is always of bounded variation if it is bounded, i.e., if $p(\mathbf{B}) < \infty$. (Henceforth it will be assumed that all nonnegative quasi-volumes are bounded.) Given two nonnegative quasi-volumes $p(B)$ and $q(B)$, defined on the same dense set of blocks Q in $\mathbf{B}$, suppose we form the difference

$$\sigma(B) = p(B) - q(B).$$

Then $\sigma(B)$, like $p(B)$ and $q(B)$, is obviously additive in the sense of Sec. 4.2, and is hence a quasi-volume. Moreover, the quasi-volume $\sigma(B)$, which is in general signed, is of bounded variation. In fact, given any set of disjoint blocks $B_1, \ldots, B_m$, we have

$$\sum_{j=1}^{m} |\sigma(B_j)| \leqslant \sum_{j=1}^{m} p(B_j) + \sum_{j=1}^{m} q(B_j) \leqslant p(\mathbf{B}) + q(\mathbf{B}),$$

so that the sum on the left is bounded by a fixed constant, as required. We now show that the converse is also true:

THEOREM 7. *Every signed quasi-volume σ of bounded variation, defined on a dense set of blocks Q in $\mathbf{B}$, can be represented as the difference between two nonnegative quasi-volumes, defined on the same set Q.*

Proof. Given any block $B \in Q$, the quantity

$$p(B) = \sup \sum_{j=1}^{m} \sigma(B_j),$$

where the least upper bound is taken with respect to all sets of disjoint subblocks $B_j \subset B$ $(j = 1, \ldots, m)$,[9] is defined and nonnegative. Moreover, it is easy to see that $p(B)$, like $\sigma(B)$, is additive, and hence a quasi-volume. In fact, let $B^{(1)}, \ldots, B^{(s)}$ be any set of disjoint blocks whose union is B, and let $B_1, \ldots, B_m$ be any set of disjoint blocks contained in B. Then, on the one hand, we have

$$\sum_{j=1}^{m} \sigma(B_j) = \sum_{j=1}^{m} \sum_{k=1}^{s} \sigma(B_j B^{(k)}) = \sum_{k=1}^{s} \sum_{j=1}^{m} \sigma(B_j B^{(k)}) \leqslant \sum_{k=1}^{s} p(B^{(k)}),$$

[9] Here, as elsewhere in Sec. 4.7, all blocks are assumed to belong to the underlying dense set Q.

which implies

$$p(B) = \sup \sum_{j=1}^{m} \sigma(B_j) \leqslant \sum_{k=1}^{s} p(B^{(k)}). \tag{30}$$

On the other hand, given any $\varepsilon > 0$, we can find a set of disjoint subblocks $B_j^{(k)}$ $(j = 1, \ldots, r_k)$ of the block $B^{(k)}$ such that

$$p(B^{(k)}) \leqslant \sum_{j=1}^{r_k} \sigma(B_j^{(k)}) + \frac{\varepsilon}{s}.$$

Summing over the index k, we obtain

$$\sum_{k=1}^{s} p(B^{(k)}) \leqslant \sum_{k=1}^{s} \sum_{j=1}^{r_k} (B_j^{(k)}) + \varepsilon \leqslant p(B) + \varepsilon,$$

and hence

$$\sum_{k=1}^{s} p(B^{(k)}) \leqslant p(B), \tag{31}$$

since ε is arbitrary. Together, (30) and (31) imply

$$\sum_{k=1}^{s} p(B^{(k)}) = p(B),$$

i.e., the function $p(B)$ is additive, and hence a quasi-volume, as asserted.

Finally let

$$q(B) = p(B) - \sigma(B). \tag{32}$$

Since $\sigma(B)$ and $p(B)$ are quasi-volumes, so is $q(B)$. Moreover, $q(B)$ is nonnegative, since $p(B) \geqslant \sigma(B)$. The theorem now follows from the formula

$$\sigma(B) = p(B) - q(B), \tag{33}$$

equivalent to (32).

4.7.2. Other representations of σ. The canonical representation. The representation (34) in terms of the nonnegative quasi-volumes p and q is not unique. In fact, let r be any nonnegative quasi-volume defined on the blocks $B \in Q$. Then, besides the representation (34), we can also write

$$\sigma(B) = [p(B) + r(B)] - [q(B) + r(B)]. \tag{34}$$

It turns out that (34) is actually the most general representation of the signed quasi-volume σ as a difference between two nonnegative quasi-volumes. To see this, suppose we have any representation

$$\sigma(B) = p_1(B) - q_1(B), \tag{35}$$

where p_1 and q_1 are nonnegative quasi-volumes. Then, given any set $B_1, \ldots, B_m$ of disjoint subblocks of B,

$$\sum_{j=1}^{m} \sigma(B_j) \leqslant \sum_{j=1}^{m} p_1(B_j) \leqslant p_1(B),$$

and hence

$$p(B) = \sup \sum_{j=1}^{n} \sigma(B_j) \leqslant p_1(B).$$

Therefore $p_1 = p + r$, where $r = p_1 - p$ is a nonnegative quasi-volume. Moreover,

$$q_1(B) = p_1(B) - \sigma(B) = p(B) - \sigma(B) + r(B) = q(B) + r(B),$$

i.e., we have reduced (35) to the form (34), as required.

At the same time, we find that the representation (33), explicitly constructed in Theorem 7, has a simple characterization in the class of all possible representations (35), i.e., the quasi-volumes p and q figuring in (33) are the *smallest possible* among all that can figure in (35). For this reason, (33) will be called the *canonical representation* of the quasi-volume σ.

4.7.3. Formulas for the positive, negative and total variations. It will be recalled that the function $p(B)$ is defined by the formula

$$p(B) = \sup \sum_{j} \sigma(B_j),$$

where the least upper bound is taken with respect to all sets of disjoint subblocks $B_j \subset B$. We now derive analogous formulas for the functions $q(B)$ and $v(B) \equiv p(B) + q(B)$.

Since the function q plays the same role in the representation of the quasi-volume

$$-\sigma(B) = q(B) - p(B)$$

as played by p in the representation of σ, and since q and p are minimal in the sense of Sec. 4.7.2, it follows that

$$q(B) = \sup \left\{ - \sum_{j} \sigma(B_j) \right\},$$

where the least upper bound is taken with respect to all sets of disjoint subblocks $B_j \subset B$. As for the quasi-volume $v(B)$, consider a set of disjoint subblocks $B_j' \subset B$ such that

$$\sum_{j} \sigma(B_j') > p(B) - \varepsilon, \tag{36}$$

and let B_j'' be a set of subblocks such that B equals the union of all the B_j' and B_j''. Then

$$\sum_{j} [-\sigma(B_j'')] = -\sigma(B) + \sum_{j} \sigma(B_j') > p(B) - \sigma(B) - \varepsilon = q(B) - \varepsilon,$$

and moreover, if B_j is any of the blocks B_j', B_j'',

$$\sum_{j} |\sigma(B_j)| = \sum_{j} |\sigma(B_j')| + \sum_{j} |\sigma(B_j'')| > p(B) + q(B) - 2\varepsilon = v(B) - 2\varepsilon.$$

Therefore

$$v(B) \leqslant \sup \sum_j |\sigma(B_j)|, \tag{37}$$

where the least upper bound is taken with respect to all partitions of the block B into disjoint subblocks.

On the other hand, if $B = \bigcup_j B_j$ is an arbitrary partition of the block B into disjoint subblocks, then

$$v(B) = \sum_j v(B_j) \geqslant \sum_j |\sigma(B_j)|, \tag{38}$$

since obviously

$$|\sigma(B)| \leqslant p(B) + q(B) = v(B).$$

Comparing (37) and (38), we find that

$$v(B) = \sup \sum_j |\sigma(B_j)|, \tag{39}$$

where the least upper bound is taken with respect to all partitions of the block B into disjoint subblocks. The quantity (39) has already been encountered in Sec. 4.2, where it was called the *total variation* of the quasi-volume σ. By the same token, the quasi-volume p is called the *positive variation* of σ, while q is called the *negative variation* of σ. Thus the total variation v is the sum of the positive and negative variations p and q.

Remark. The above construction greatly resembles that given in Sec. 2.11, where we represented a functional of variable sign as the difference between two nonnegative functionals. This suggests the possibility of a direct connection between the two constructions. Such a connection in fact exists, as we shall see in Sec. 5.6.

4.7.4. The case $n = 1$. Jordan's theorem. Consider the case $n = 1$, where the basic block $\mathbf{B}$ is a closed interval $[a, b]$. Let $\sigma(\alpha, \beta]$ be a quasi-length defined on $[a, b]$, with generating function

$$F(x) = \sigma[a, x], \qquad F(a) = 0.$$

Then, as we know from Sec. 4.3, $\sigma(\alpha, \beta]$ is of bounded variation in $[a, b]$ if and only if $F(x)$ is of bounded variation in $[a, b]$, in the sense that

$$\sup \sum_{j=1}^{m} |F(x_j) - F(x_{j-1})| < \infty, \tag{40}$$

where the least upper bound is taken with respect to all partitions

$$a = x_0 < x_1 < \cdots < x_{m-1} < x_m = b. \tag{41}$$

On p. 65 the quantity (40) was called the total variation of $F(x)$ in the interval $[a, b]$, denoted by $V_a^b(F)$. If $F(x)$ is of bounded variation in $[a, b]$, then obviously $F(x)$ is also of bounded variation in every subinterval of the form $[a, x]$, $a < x \leqslant b$, and the quantity (40), where the least upper bound is taken with respect to all partitions (41) with $x_m = x$ instead of $x_m = b$, becomes a function of x, which we still call the *total variation* of $F(x)$ and denote by $V_a^x(F)$, $\mathrm{Var}_a^x(F)$ or simply $V(x)$. We are now in a position to analyze the structure of functions of bounded variation:

THEOREM 8 (*Jordan's theorem*). *Every generating function $F(x)$ of bounded variation in the interval $[a, b]$ can be represented in the form*

$$F(x) = P(x) - Q(x),$$

where $P(x)$ and $Q(x)$ are bounded, nonnegative and nondecreasing generating functions. Moreover, the total variation of $F(x)$ is given by

$$V(x) = P(x) + Q(x).$$

Proof. The corresponding quasi-length defined by

$$\sigma(\alpha, \beta] = F(\beta) - F(\alpha)$$

can be represented in the form

$$\sigma(\alpha, \beta] = p(\alpha, \beta] - q(\alpha, \beta],$$

where p and q are (bounded) nonnegative quasi-lengths. The rest of the proof follows by setting

$$P(x) = p[a, x], \qquad Q(x) = q[a, x].$$

Remark 1. The functions $P(x)$ and $Q(x)$ are called the *positive* and *negative variations* of $F(x)$. In terms of $P(x)$ and $Q(x)$, we obviously have

$$I_\sigma f = \int_a^b f(x)\, dF(x) = \int_a^b f(x)\, dP(x) - \int_a^b f(x)\, dQ(x).$$

Remark 2. Analogous results can be deduced for arbitrary n, but they are less interesting because of the complexity of the relation between quasi-volumes and generating functions when $n > 1$ (see the remark on p. 65).

Remark 3. Jordan's theorem remains true if the adjective "generating" is omitted (twice). In fact, the class of generating functions of bounded variation (in the interval $[a, b]$) is just the subset of the class of functions of bounded variation satisfying the extra condition $F(a) = 0$. Thus, suppose $F(x)$ is a function of bounded variation such that $F(a) \neq 0$. Then $F(x) - F(a)$ is a generating function of bounded variation, and we need only add $F(a)$ to $P(x)$ or $Q(x)$, depending on whether $F(a)$ is positive or negative.

PROBLEMS

1. Evaluate the following Stieltjes integrals:

a) $I_1 = \int_0^2 x\, dF(x),$ where $F(x) = \begin{cases} x^2 & \text{for } 0 \leqslant x < 1, \\ x+1 & \text{for } 1 \leqslant x \leqslant 2; \end{cases}$

b) $I_2 = \int_{-1}^3 x\, dF(x),$ where $F(x) = \begin{cases} 0 & \text{for } x = -1, \\ 1 & \text{for } -1 < x < 2, \\ -1 & \text{for } 2 \leqslant x \leqslant 3. \end{cases}$

Ans. a) $I_1 = \frac{19}{6}$; b) $I_2 = -5$.

2 (*The Cantor function*). Consider the function $C(x)$, $0 \leqslant x \leqslant 1$, defined as follows: At every point x of the Cantor set C (see Prob. 2, p. 21) with a ternary expansion $x = 0.\theta_1\theta_2\ldots$ (the numbers θ_n take the values 0 or 2), the function $C(x)$ has a binary expansion $0.\theta_1'\theta_2'\ldots$, where $\theta_n' = \theta_n/2$. Then $C(x)$ takes equal values at the end points of every interval $[\alpha, \beta]$ adjacent to C (in the sense of Prob. 3, p. 22), and the definition of $C(x)$ is completed by setting $C(x)$ equal to the corresponding constant in the whole interval $[\alpha, \beta]$. The function $C(x)$ is called the *Cantor function*. Show that $C(x)$ is continuous.

Hint. $C(x)$ is nondecreasing, and its range is dense in $[0, 1]$.

3. Show that the Cantor function $C(x)$ defined in the preceding problem cannot be represented in the form

$$C(x) = \int_0^x g(x)\, dx, \tag{42}$$

where $g(x)$ is a summable function.

Hint. Assuming that (42) holds, show that $g(x)$ must vanish at almost every point of the complement of the Cantor set, thereby establishing a contradiction.

Comment. The Cantor function is the generating function of a quasi-length which does not reduce to either of the types considered in Examples 2 and 3, p. 64.

4. Show that the product of two functions $F_1(x)$ and $F_2(x)$ of bounded variation is also of bounded variation, where

$$V_a^b(F_1F_2) \leqslant \max|F_1(x)|\, V_a^b(F_2) + \max|F_2(x)|\, V_a^b(F_1).$$

5. Let $F(x) \geqslant \alpha > 0$ be a function of bounded variation. Show that $1/F(x)$ is also a function of bounded variation, where

$$V_a^b\left(\frac{1}{F}\right) \leqslant \frac{1}{\alpha^2} V_a^b(F).$$

6. A curve $y = F(x)$, $a \leqslant x \leqslant b$ is said to be *rectifiable* if the length of the "inscribed polygonal curve" with consecutive vertices at the points $(x_0, F(x_0))$, $(x_1, F(x_1)), \ldots, (x_{n-1}, F(x_{n-1}))$, $(x_n, F(x_n))$, where

$$a = x_0 < x_1 < \cdots < x_{n-1} < x_n = b,$$

is bounded by a fixed constant independent of n and the choice of the intermediate points $x_1, \ldots, x_{n-1}$. Prove that the curve $y = F(x)$ is rectifiable if and only if the function $F(x)$ is of bounded variation.

Hint. Use the inequality

$$|\Delta y_j| \leqslant \sqrt{(\Delta x_j)^2 + (\Delta y_j)^2} \leqslant |\Delta x_j| + |\Delta y_j|.$$

7. Show that the continuous function

$$x^\alpha \sin \frac{1}{x^\beta} \qquad (0 \leqslant x \leqslant 1;\ \alpha, \beta > 0)$$

is of bounded variation if $\alpha > \beta$, but not if $\alpha \leqslant \beta$.

8. Let $\mathscr{B}$ be the space of all functions $F(x)$ of bounded variation in the interval $[a, b]$, where functions differing by a constant are regarded as equivalent. Show that $\mathscr{B}$ is a complete normed linear space, when equipped with the norm

$$\|F\| = V_a^b(F).$$

5

THE LEBESGUE-STIELTJES INTEGRAL

5.1. Definition of the Lebesgue-Stieltjes Integral

The natural next step is to construct a Lebesgue-Stieltjes integral which generalizes the Riemann-Stieltjes integral in the same way as the Lebesgue integral generalizes the ordinary Riemann integral. As shown in Chap. 3, the Lebesgue integral can be constructed starting from either step functions or continuous functions as the elementary functions, depending on our preference. However, these two constructions are no longer equivalent in the case of the Lebesgue-Stieltjes integral. More exactly, we can always construct the Lebesgue-Stieltjes integral starting from continuous functions as the elementary functions, with the Riemann-Stieltjes integral as the elementary integral, but if step functions are chosen as the elementary functions, the construction is possible only when extra conditions are imposed on the original quasi-volume (see Sec. 5.8). For this reason, we prefer to begin with the first approach, using continuous functions as the elementary functions and the Riemann-Stieltjes integral as the elementary integral.

Let σ be a quasi-volume defined on a dense set of blocks Q in the basic block $\mathbf{B}$. For the time being, we assume that σ is nonnegative, i.e., that $\sigma(B) \geqslant 0$ for every block $B \in Q$. Let $\tilde{H}$ be the set of all functions continuous in $\mathbf{B}$. Then $\tilde{H}$ obviously satisfies the conditions on p. 23 for a family of elementary functions, with $\mathbf{B}$ playing the role of the set X. Moreover, if we define the elementary integral of any function $f \in \tilde{H}$ as the Riemann-Stieltjes integral

$$\tilde{I}_\sigma f = \int_{\mathbf{B}} f(x)\sigma(dx),$$

then it is easy to see that $\tilde{I}_\sigma f$ satisfies the axioms on p. 24 for an elementary integral. In fact, Axioms 1 and 2 are obvious, while Axiom 3 is an immediate consequence of the estimate

$$|\tilde{I}_\sigma f_m| = \left| \int_{\mathbf{B}} f_m(x)\sigma(dx) \right| \leqslant V_{\mathbf{B}}(\sigma) \max_{x \in \mathbf{B}} |f_m(x)|$$

[cf. formula (12), p. 68] and Dini's lemma, just as in Sec. 3.4.

We are now in a position to apply the general scheme of Chap. 2, obtaining first a class $\tilde{L}_\sigma^+(\mathbf{B})$ of functions f which are (almost-everywhere) limits of nondecreasing sequences $h_m \in \tilde{H}$ with bounded Riemann-Stieltjes integrals,[1] and then a class $\tilde{L}_\sigma(\mathbf{B})$ of functions φ which are differences of functions in $\tilde{L}_\sigma^+(\mathbf{B})$. The functions in $\tilde{L}_\sigma$ are said to be *σ-summable* or *Lebesgue-Stieltjes integrable* (with respect to the quasi-volume σ), and the corresponding Lebesgue-Stieltjes integral is denoted by $\tilde{I}_\sigma \varphi$ or

$$\int_{\mathbf{B}} \varphi(x)\sigma(dx).$$

Moreover, $\tilde{L}_\sigma$ is a complete normed linear space, when equipped with the norm

$$\|\varphi\|_\sigma = \tilde{I}_\sigma(|\varphi|).$$

In the case $n = 1$, the Lebesgue-Stieltjes integral over the closed interval $[a, b]$ is denoted by $\tilde{I}_\sigma \varphi$,

$$\int_a^b \varphi(x)\sigma(dx),$$

or

$$\int_a^b \varphi(x)\, dF(x),$$

where $F(x)$ is the generating function of the quasi-length $\sigma(\alpha, \beta]$.

5.2. Examples

Naturally, the character of $\tilde{L}_\sigma$ depends on the quasi-volume σ, as we now illustrate, using the same examples as in Sec. 4.2.

Example 1. Let

$$\sigma(B) = s(B),$$

where $s(B)$ is the volume of the block B in the ordinary sense (and $\mathbf{B}$ is finite). Then obviously $\tilde{L}_\sigma$ is just the class of Lebesgue-integrable functions.

---

[1] Here convergence almost everywhere is defined in the obvious way, i.e., a set $Z \subset X$ is called a set of *σ-measure zero* (relative to the elementary integral $\tilde{I}$) if given any $\varepsilon > 0$, there exists a nondecreasing sequence of nonnegative functions $h_p(x) \in \tilde{H}$ such that $\tilde{I}_\sigma h_p < \varepsilon$ and $\sup h_p(x) \geqslant 1$ on Z, and a sequence is said to converge almost everywhere if it converges everywhere except on a set of σ-measure zero.

Example 2. Let

$$\sigma(B) = \int_B g(x)\, dx,$$

where $g(x) \geqslant 0$ is a function summable over $\mathbf{B}$, and let $h(x)$ be a function continuous in $\mathbf{B}$. Then every Riemann-Stieltjes sum of $h(x)$, with respect to the quasi-volume σ, is of the form

$$\sum_{j=1}^{m} h(\xi_j)\sigma(B_j) = \sum_{j=1}^{m} h(\xi_j)\int_{B_j} g(x)\, dx$$
$$= \sum_{j=1}^{m} \int_{B_j} h(\xi_j) g(x)\, dx = \int_{\mathbf{B}} h_\Pi(x) g(x)\, dx,$$

where $h_\Pi(x)$ is the step function equal to $h(\xi_j)$ in the block B_j $(j = 1, \ldots, m)$. Since $h_\Pi(x)$ converges uniformly to the function $h(x)$ as the partition is refined indefinitely, i.e., as $d(\Pi) \to 0$, it follows from Lebesgue's theorem (see Sec. 2.7) that

$$\tilde{I}_\sigma h = \int_{\mathbf{B}} h(x)\sigma(dx) = \lim_{d(\Pi)\to 0} \sum_{j=1}^{m} h(\xi_j)\sigma(B_j) = \int_{\mathbf{B}} h(x) g(x)\, dx = I(hg),$$

where I is the ordinary Lebesgue integral.

Now let h_p be a nondecreasing sequence of elementary functions which converges (everywhere) to a function f, and suppose the integrals $\tilde{I}_\sigma h_p$, and hence the Lebesgue integrals $I(h_p g)$, form a bounded sequence. Then the limit fg of the sequence $h_p g$ is summable in the ordinary sense, and

$$\tilde{I}_\sigma f = \lim_{p\to\infty} \tilde{I}_\sigma h_p = \lim_{p\to\infty} I(h_p g) = I(fg).$$

Such a function belongs to the class $\tilde{L}_\sigma^+$ by definition. Moreover, any function in $\tilde{L}_\sigma^+$ differs from a function of this type only by a function f_0 which vanishes everywhere except on a set of σ-measure zero. It follows that $I(f_0 g) = 0$. In fact, let $Z = \{x : f_0(x) \neq 0\}$. Since Z is a set of σ-measure zero, given any positive integer m, there exists a nondecreasing sequence of nonnegative continuous functions $h_p^{(m)}(x)$ such that $\tilde{I}_\sigma h_p^{(m)} < 1/m$ and $\sup_p h_p^{(m)}(x) \geqslant 1$ on Z. Moreover, it can be assumed that

$$h_p^{(m+1)}(x) \leqslant h_p^{(m)}(x)$$

for arbitrary p and m. For fixed m, $h_p^{(m)}$ is a nondecreasing sequence converging to some limit $h^{(m)}(x)$. Taking the limit as $m \to \infty$ of the nonincreasing sequence $h^{(m)}(x)$, we obtain a function $h(x)$ which is $\geqslant 1$ on Z. According to Levi's theorem,

$$I(h^{(m)} g) = \lim_{p\to\infty} I(h_p^{(m)} g) = \lim_{p\to\infty} \tilde{I}_\sigma(h_p^{(m)}),$$

and hence for any m we have

$$I(h^{(m)} g) \leqslant \frac{1}{m}.$$

Therefore, by Levi's theorem again,

$$I(hg) = \lim_{m\to\infty} I(h^{(m)}g) = 0.$$

Therefore the function hg vanishes almost everywhere (relative to the ordinary Lebesgue integral I). But then f_0g also vanishes almost everywhere, since if $f(x_0)g(x_0) \neq 0$ for some x_0, then $f_0(x_0) \neq 0$ and $g(x_0) \neq 0$, i.e., $h(x_0) \geqslant 1$, $g(x_0) \neq 0$ and $h(x_0)g(x_0) \neq 0$. Therefore $I(f_0g) = 0$, as asserted.

Thus we have finally shown that if $f \in \tilde{L}_\sigma^+$, then the product fg is summable (in the ordinary sense) and

$$\tilde{I}_\sigma f = I(fg).$$

Taking differences, we see that *if* φ *is* σ*-summable* ($\varphi \in \tilde{L}_\sigma$), *then the product* φg *is summable*, and moreover

$$\tilde{I}_\sigma \varphi = I(\varphi g). \tag{1}$$

The converse is also true, i.e., *if the product* φg *is summable, then* φ *is* σ*-summable*. However, the proof of this assertion requires a deeper knowledge of measure theory, and hence will be postponed until Sec. 7.6.

Example 3. Given a sequence of points $c_1, \ldots, c_m, \ldots$ in the basic block $\mathbf{B}$ and a sequence of *positive* real numbers $g_1, \ldots, g_m, \ldots$ such that

$$\sum_{m=1}^{\infty} g_m < \infty,$$

let $\sigma(B)$ equal the sum of all the g_m such that the corresponding points c_m belong to the block $\mathbf{B}$. Moreover, let the function $h(x)$ be continuous in $\mathbf{B}$, and let $\Pi = \{B_j\}$ be a partition of $\mathbf{B}$. Then

$$\begin{aligned} I_\sigma h &= \int_{\mathbf{B}} h(x)\sigma(dx) = \lim_{d(\Pi)\to 0} \sum_{j=1}^{m} h(\xi_j)\sigma(B_j) = \lim_{d(\Pi)\to 0} \sum_{j=1}^{m} h(\xi_j) \sum_{c_k \in B_j} g_k \\ &= \sum_{j=1}^{m} \sum_{c_k \in B_j} h(c_k)g_k + \lim_{d(\Pi)\to 0} \sup_{\substack{\xi_j, c_k \in B_j \\ B_j \in \Pi}} |h(\xi_j) - h(c_k)| \sum_{k=1}^{\infty} g_k \\ &= \sum_{k=1}^{\infty} h(c_k)g_k, \end{aligned} \tag{2}$$

since $h(x)$ is uniformly continuous in $\mathbf{B}$.

Now let h_p be a nondecreasing sequence of elementary (i.e., continuous) functions which converges to a function f at every point c_k, and suppose the integrals $\tilde{I}_\sigma h_p$ form a bounded sequence. Then, as we know, the limit function f belongs to the class $\tilde{L}_\sigma^+$. It can be assumed in this argument that the functions h_p are nonnegative, since otherwise we need only replace h_p by $h_p - h_1$ and f by $f - h_1$. Using (2), we have

$$\tilde{I}_\sigma h_p = \sum_{k=1}^{\infty} h_p(c_k)g_k \leqslant C,$$

and hence

$$h_p(c_k)g_k \leqslant C$$

for all p. It follows that $h_p(c_k) \nearrow f(c_k)$ for every fixed k. Moreover, for every N,

$$C \geqslant \sum_{k=1}^{N} h_p(c_k)g_k \nearrow \sum_{k=1}^{N} f(c_k)g_k,$$

which implies

$$\sum_{k=1}^{\infty} f(c_k)g_k = \lim_{N\to\infty} \sum_{k=1}^{N} f(c_k)g_k \leqslant C.$$

Finally, given any $\varepsilon > 0$, we have

$$\sum_{k=N}^{\infty} f(c_k)g_k < \varepsilon$$

for sufficiently large N. Therefore

$$0 \leqslant \sum_{k=1}^{\infty} f(c_k)g_k - \sum_{k=1}^{\infty} h_p(c_k)g_k \leqslant \left[\sum_{k=1}^{N-1} f(c_k)g_k - \sum_{k=1}^{N-1} h_p(c_k)g_k\right] + \sum_{k=N}^{\infty} f(c_k)g_k < 2\varepsilon$$

for sufficiently large p, and hence

$$\tilde{I}_\sigma f = \lim_{p\to\infty} \tilde{I}_\sigma h_p = \sum_{k=1}^{\infty} f(c_k)g_k.$$

Thus we have shown that if $f \in \tilde{L}_\sigma^+$, then the series

$$\sum_{k=1}^{\infty} f(c_k)g_k$$

converges and equals the integral $\tilde{I}_\sigma f$. It should be noted that the values of $f(x)$ at points of the block **B** other than c_k play no role at all, and in fact, *$f(x)$ need only be defined at the points c_k*. Taking differences, we see that *if φ is defined at every point c_k and if $\varphi \in \tilde{L}_\sigma$, then the series*

$$\sum_{k=1}^{\infty} \varphi(c_k)g_k$$

converges and equals the integral $\tilde{I}_\sigma\varphi$, and moreover

$$\sum_{k=1}^{\infty} |\varphi(c_k)|g_k < \infty. \tag{3}$$

Conversely, *if (3) holds, then φ belongs to $\tilde{L}_\sigma$*. In fact, suppose φ is nonzero at only one of the points c_k, say c_1. Then, if $\varphi(c_1)$ is positive, φ is the limit of a nonincreasing sequence of continuous functions $h_p(x)$ equal to $\varphi(c_1)$ at the point c_1.[2] Therefore, in this case, $\varphi \in \tilde{L}_\sigma$ (in fact, $\varphi \in \tilde{L}_\sigma^+$), and $\tilde{I}_\sigma\varphi$ is just $\varphi(c_1)g_1$. If $\varphi(c_1)$ is negative, the same argument can be used to show that $-\varphi$ belongs to $\tilde{L}_\sigma^+$, so that φ again belongs to $\tilde{L}_\sigma$, with the same integral as before. In the general case, φ is a sum of functions, each "concentrated" at one of the points c_k, where the sum of the corresponding integrals is absolutely convergent, by hypothesis. Therefore φ belongs to $\tilde{L}_\sigma$, by an obvious version of Levi's theorem. Our description of $\tilde{L}_\sigma$ is now complete.

Remark. In each of the above examples, the Lebesgue-Stieltjes integral of φ turns out to be a numerical series or the ordinary Lebesgue integral of the product of φ with some function. In the general case, the Lebesgue-Stieltjes integral has a more complicated structure (cf. Prob. 3, p. 86).

5.3. The Lebesgue-Stieltjes Integral with Respect to a Signed Quasi-Volume

As we know from Sec. 4.7, a signed quasi-volume σ (of bounded variation) can be represented as the difference between two nonnegative quasi-volumes p and q, called the positive and negative variations of σ. Let v be the quasi-volume $p + q$, i.e., the total variation of σ. Then, using the nonnegative quasi-volumes v, p and q, we can construct corresponding spaces $\tilde{L}_v$, $\tilde{L}_p$ and $\tilde{L}_q$ of summable functions. Every function $\varphi \in \tilde{L}_v$ also belongs to the spaces $\tilde{L}_p$ and $\tilde{L}_q$, as we shall prove in a moment. Thus, if $\varphi \in \tilde{L}_v$, the integrals $\tilde{I}_v\varphi$, $\tilde{I}_p\varphi$ and $\tilde{I}_q\varphi$ all exist, where, as is easily verified,

$$\tilde{I}_v\varphi = \tilde{I}_p\varphi + \tilde{I}_q\varphi.$$

This suggests the following

DEFINITION. *If $\varphi \in \tilde{L}_v$, the Lebesgue-Stieltjes integral of φ is given by*

$$\tilde{I}_\sigma\varphi = \tilde{I}_p\varphi - \tilde{I}_q\varphi. \tag{4}$$

For continuous φ, the expression (4) coincides with the Riemann-Stieltjes integral of φ with respect to σ. Moreover, as shown in Sec. 4.7.2, the canonical representation $\sigma = p - q$ is the "most economical," in the sense that $p_1 \geqslant p$, $q_1 \geqslant q$, $v_1 = p_1 + q_1 \geqslant v$ for any other

[2] In the one-dimensional case, choose the functions

$$h_p(x) = \begin{cases} \varphi(c_1)\{1 - p\,|x - c_1|\} & \text{for} \quad |x - c_1| \leqslant 1/p,\ x \in \mathbf{B}, \\ 0 & \text{otherwise.} \end{cases}$$

In the n-dimensional case, represent $h_p(x)$ as a product of n such functions, each depending on one coordinate.

representation. Therefore the space $\tilde{L}_v$ (corresponding to the canonical representation) is the largest possible ($\tilde{L}_{v_1} \subset \tilde{L}_v$).

We now establish the missing step, as promised. Let p and v be any two nonnegative quasi-volumes, defined on some dense set of blocks Q in $\mathbf{B}$, such that $p(B) \leqslant v(B)$ for every $B \in Q$. Then $\tilde{L}_v \subset \tilde{L}_p$, and moreover

$$\tilde{I}_p(|\varphi|) \leqslant \tilde{I}_v(|\varphi|) \tag{5}$$

for every $\varphi \in \tilde{L}_v$. For continuous φ, this is obvious, since then

$$\tilde{I}_p(|\varphi|) = \int_{\mathbf{B}} |\varphi(x)| \, p(dx) \leqslant \int_{\mathbf{B}} |\varphi(x)| \, v(dx) = \tilde{I}_v(|\varphi|).$$

More generally, suppose φ is the limit (everywhere) of a nonincreasing sequence of continuous functions h_m, with bounded integrals $\tilde{I}_v h_m$. Then the integrals $\tilde{I}_p h_m$ are also bounded, so that the function φ belongs to $\tilde{L}_p^+$. However, φ is still not sufficient for our purposes, since the general function in $\tilde{L}_v^+$ differs from a function like φ by a function φ_0 vanishing almost everywhere relative to $\tilde{I}_v$. But φ_0 also vanishes almost everywhere relative to $\tilde{I}_p$. In fact, given any $\varepsilon > 0$, there exists a nondecreasing sequence of nonnegative continuous functions $h_m(x)$ such that $\tilde{I}_v h_m < \varepsilon$ and $\sup h_m(x) \geqslant 1$ on the set $Z = \{x \colon \varphi_0(x) \neq 0\}$. Therefore, since $\tilde{I}_p h_m \leqslant \tilde{I}_v h_m$, the set Z is also of measure zero relative to $\tilde{I}_p$. It follows that φ_0 belongs to $\tilde{L}_p$, and hence so does the general function in $\tilde{L}_v^+$. Moreover, any function $\psi \in \tilde{L}_v$ belongs to $\tilde{L}_p$, being the difference between two functions in $\tilde{L}_v^+$, i.e., $\tilde{L}_v \subset \tilde{L}_p$, as required. Finally, to see that the inequality (5), valid for continuous φ, continues to hold for every $\varphi \in \tilde{L}_v$, we need only take the limit with respect to the $\tilde{L}_v$-norm, noting that the set of elementary functions is dense in the space of summable functions (cf. Sec. 2.9).

5.4. The General Continuous Linear Functional on the Space $C(\mathbf{B})$

Given a (signed) quasi-volume σ of bounded variation, we can form the Riemann-Stieltjes integral

$$I_\sigma f = \int_{\mathbf{B}} f(x) \sigma(dx)$$

of any function $f(x)$ continuous in the basic block $\mathbf{B}$. Let $C(\mathbf{B})$ be the normed linear space of all functions continuous in $\mathbf{B}$, equipped with the norm

$$\|f\| = \max_{x \in \mathbf{B}} |f(x)|.$$

Then the integral $I_\sigma f$ defines a continuous linear functional on $C(\mathbf{B})$, since it satisfies the following two conditions:

1) If f_1, f_2 are any two functions in $C(\mathbf{B})$ and α_1, α_2 are any two real numbers, then
$$I_\sigma(\alpha_1 f_1 + \alpha_2 f_2) = \alpha_1 I_\sigma f_1 + \alpha_1 I_\sigma f_2.$$

2) If $f_m \in C(\mathbf{B})$ is a sequence such that $\|f_m\| \to 0$ as $m \to \infty$, then $I_\sigma f_m \to 0$, as follows at once from the estimate
$$|I_\sigma f_m| = \left| \int_{\mathbf{B}} f_m(x)\sigma(dx) \right| \leqslant \|f_m\| V_{\mathbf{B}}(\sigma)$$
(cf. p. 89).

Thus every Riemann-Stieltjes integral gives rise to a continuous linear functional on $C(\mathbf{B})$. We now prove the converse:

THEOREM 1. *Given a continuous linear functional If defined on the space* $C(\mathbf{B})$, *there exists a quasi-volume* $\tilde{\sigma} = \tilde{\sigma}(I)$ *of bounded variation such that*
$$If = \int_{\mathbf{B}} f(x)\tilde{\sigma}(dx) \tag{6}$$
for every $f \in C(\mathbf{B})$.

Proof. First suppose I is nonnegative, so that $If \geqslant 0$ if $f(x) \geqslant 0$. Then, choosing $C(\mathbf{B})$ as the space of elementary functions and the functional I as the elementary integral, we can construct a space L_I of I-summable functions. The only nontrivial part of this assertion, given the theory of Chap. 2, is to verify that I satisfies Axiom 3, p. 24. But, according to Dini's lemma (p. 54), $f_m \searrow 0$ implies $f_m \to 0$ uniformly, i.e., $\|f_m\| \to 0$, and hence $If_m \to 0$. In particular, L_I contains the characteristic function
$$\chi_B(x) = \begin{cases} 1 & \text{for} \quad x \in B, \\ 0 & \text{for} \quad x \notin B \end{cases}$$
of every block $B \subset \mathbf{B}$. In fact, $\chi_B(x)$ can be represented (in various ways) as the limit of an everywhere convergent sequence of continuous functions $f_m(x)$, where the functions $f_m(x)$ can be chosen to be nonnegative and bounded by 1 (the construction resembles that given in footnote 2, p. 93). Therefore
$$If_m \leqslant I(1),$$
and hence, by Lebesgue's theorem (see Sec. 2.7), $\chi_B \in L_I$ and
$$I\chi_B = \lim_{m \to \infty} If_m.$$
Defining
$$\tilde{\sigma}(B) = I\chi_B, \tag{7}$$

we see that $\tilde{\sigma}(B)$ is bounded, since $\tilde{\sigma}(B) \leqslant \tilde{\sigma}(\mathbf{B}) = I(1)$, and obviously represents a nonnegative quasi-volume defined on *every* block $B \subset \mathbf{B}$. Moreover, the quasi-volume (7) satisfies the relation (6). To see this, note that the Riemann-Stieltjes integral in the right-hand side of (6) is the limit of the sum

$$\sum_{j=1}^{m} f(\xi_j)\tilde{\sigma}(B_j) = \sum_{j=1}^{m} f(\xi_j)I\chi_{B_j} = I\left\{\sum_{j=1}^{m} f(\xi_j)\chi_{B_j}(x)\right\} = Ih_\Pi(x)$$

as the partition $\Pi = \{B_j\}$ is refined indefinitely, where $h_\Pi(x)$ is the step function equal to $f(\xi_j)$ in the block B_j $(j = 1, \ldots, m)$. As $d(\Pi) \to 0$, $h_\Pi(x)$ converges uniformly to $f(x)$, and hence, by Lebesgue's theorem again,

$$\lim_{d(\Pi)\to 0} h_\Pi(x) = If,$$

as asserted.

Finally, if the functional I takes values of either sign, then, according to Riesz's representation theorem (p. 44), we can represent I in the form

$$I = J - N,$$

where the linear functionals J and N are nonnegative and continuous in the sense that $h_m \searrow 0$ implies $Jh_m \to 0$, $Nh_m \to 0$. Let $\tilde{p}$ and $\tilde{q}$ be the nonnegative quasi-volumes corresponding to J and N, in accordance with the above construction. Then, given any function $f \in C(\mathbf{B})$,

$$If = Jf - Nf = \int_{\mathbf{B}} f(x)\tilde{p}(dx) - \int_{\mathbf{B}} f(x)\tilde{q}(dx) \equiv \int_{\mathbf{B}} f(x)\tilde{\sigma}(dx),$$

and If can once again be written in the form (6), where $\tilde{\sigma}$ is a quasi-volume of bounded variation. This completes the proof.

Remark. Thus, without knowing it at the time, in carrying out the constructions of Secs. 5.1 and 5.3, we were actually using the most general continuous linear functional as the elementary integral!

5.5. Relation between the Quasi-Volumes σ and $\tilde{\sigma}$

Suppose the functional If figuring in Theorem 1 is itself a Riemann-Stieltjes integral, with respect to a quasi-volume σ defined on a dense set of blocks Q in the basic block

$$\mathbf{B} = \{x: a_1 \leqslant x_1 \leqslant b_1, \ldots, a_n \leqslant x_n \leqslant b_n\}.$$

Then it might be expected at first that

$$\tilde{\sigma}(B) = \tilde{I}_\sigma \chi_B$$

would coincide with $\sigma(B)$ for every $B \in Q$. Nevertheless, in general this is not the case (see Prob. 7, p. 109), and it can only be said that

$$\tilde{\tilde{\sigma}} = \tilde{\sigma}, \tag{8}$$

i.e., that repetition of the process leading from σ to $\tilde{\sigma}$ gives nothing new. In fact, if the sequence f_m has the same meaning as on p. 95, then obviously

$$\tilde{\tilde{\sigma}}(B) = \lim_{m \to \infty} I_{\tilde{\sigma}} f_m = \lim_{m \to \infty} I_\sigma f_m = \tilde{\sigma}(B),$$

as asserted. Note, however, that $\tilde{\sigma}(\mathbf{B})$ always coincides with $\sigma(\mathbf{B})$, since $f_m(x) \equiv 1$ [i.e., $f_m(x) = 1$ for all x in the basic block $\mathbf{B}$] is a sequence of continuous functions converging (trivially) to $\chi_{\mathbf{B}}$, and obviously in this case

$$I_\sigma f_m = \int_{\mathbf{B}} \sigma(dx) = \sigma(\mathbf{B}). \qquad (m = 1, 2, \ldots).$$

To find a direct connection between σ and $\tilde{\sigma}$, we must first introduce some new concepts. The block

$$B_1 = \{x : \alpha_1^{(1)} < x_1 \leqslant \beta_1^{(1)}, \ldots, \alpha_n^{(1)} < x_n \leqslant \beta_n^{(1)}\}$$

is said to be *strictly included* in the block

$$B_2 = \{x : \alpha_1^{(2)} < x_1 \leqslant \beta_1^{(2)}, \ldots, \alpha_n^{(2)} < x_n \leqslant \beta_n^{(2)}\}$$

(symbolically $B_1 \Subset B_2$ or $B_2 \Supset B_1$) if

1) $B_1 \subset B_2$ in the usual sense, i.e.,

$$\alpha_k^{(2)} \leqslant \alpha_k^{(1)} < \beta_k^{(1)} \leqslant \beta_k^{(2)} \qquad (k = 1, \ldots, n);$$

2) For all k,

$$\alpha_k^{(2)} < \alpha_k^{(1)} \qquad \text{if} \quad \alpha_k^{(2)} \neq a_k,$$

$$\beta_k^{(1)} < \beta_k^{(2)} \qquad \text{if} \quad \beta_k^{(2)} \neq b_k,$$

but if $\alpha_k^{(2)} = a_k$, the relation $\alpha_k^{(1)} = \alpha_k^{(2)} = a_k$ is permitted, and similarly for the "upper coordinates" $\beta_k^{(1)}$ and $\beta_k^{(2)}$.

If $B_1 \Subset B_2$, there exists a continuous function $f(x)$ in $\mathbf{B}$, taking values between 0 and 1, such that

$$f(x) = \begin{cases} 1 & \text{for} \quad x \in B_1, \\ 0 & \text{for} \quad x \notin B_2. \end{cases} \tag{9}$$

For the case $n = 1$, this is obviously true for a suitable "trapezoidal function" $f(x)$, and for arbitrary n, we need only form a product of n such functions, each depending on one coordinate (cf. footnote 2, p. 93). If $B_1 \Subset B_2$, then it follows at once from the definition of the Riemann-Stieltjes integral that

$$\sigma(B_1) \leqslant \int_{\mathbf{B}} f(x)\sigma(dx) \leqslant \sigma(B_2)$$

for the function (9) and any nonnegative quasi-volume σ.

We still need another definition. The sequence of blocks

$$B_m = \{x: \alpha_1^{(m)} < x_1 \leqslant \beta_1^{(m)}, \ldots, \alpha_n^{(m)} < x_n \leqslant \beta_n^{(m)}\}$$

is said to *converge downward* to the block

$$B = \{x: \alpha_1 < x \leqslant \beta_1, \ldots, \alpha_n < x_n \leqslant \beta_n\}$$

(symbolically $B_m \searrow B$), if for all k, $\alpha_k^{(m)} \searrow \alpha_k$, $\beta_k^{(m)} \searrow \beta_k$, where

$$\begin{aligned} \alpha_k < \alpha_k^{(m)} < \beta_k < \beta_k^{(m)} \quad &\text{if} \quad \alpha_k \neq a_k, \beta_k \neq b_k, \\ \alpha_k = \alpha_k^{(m)} < \beta_k < \beta_k^{(m)} \quad &\text{if} \quad \alpha_k = a_k, \\ \alpha_k < \alpha_k^{(m)} < \beta_k = \beta_k^{(m)} \quad &\text{if} \quad \beta_k = b_k. \end{aligned}$$

We are now in a position to find the relation between σ and $\tilde{\sigma}$. Given a sequence of blocks $B_m \searrow B$, we construct two other sequences $B'_m \searrow B$ and $B''_m \searrow B$ such that $B'_m \Supset B_m \Supset B''_m$. Let $f_m(x)$ and $g_m(x)$ be two sequences of continuous functions, taking values between 0 and 1, such that

$$f_m(x) = \begin{cases} 1 & \text{for} \quad x \in B''_m, \\ 0 & \text{for} \quad x \notin B_m, \end{cases}$$

$$g_m(x) = \begin{cases} 1 & \text{for} \quad x \in B_m, \\ 0 & \text{for} \quad x \notin B'_m. \end{cases}$$

Then

$$\int_{\mathbf{B}} f_m(x)\sigma(dx) \leqslant \sigma(B_m) \leqslant \int_{\mathbf{B}} g_m(x)\sigma(dx) \tag{10}$$

for any nonnegative quasi-volume σ. But as $m \to \infty$, both $f_m(x)$ and $g_m(x)$ converge to the characteristic function of the block B, and hence

$$\tilde{\sigma}(B) = \lim_{m\to\infty} \tilde{I}_\sigma f_m(x) = \lim_{m\to\infty} \tilde{I}_\sigma g_m(x). \tag{11}$$

Comparing (10) and (11), we see at once that the limit

$$\tilde{\sigma}(B) = \lim_{m\to\infty} \sigma(B_m) \tag{12}$$

is uniquely defined for any nonnegative quasi-volume σ and any sequence of blocks $B_m \in Q(\sigma)$ converging downward to the (arbitrary) block $B \subset \mathbf{B}$.

Formula (12) gives the desired relation between the quasi-volume $\tilde{\sigma}$ constructed via Theorem 1 and the original quasi-volume σ, for the case where σ is nonnegative. In the general case, where σ is a signed quasi-volume (of bounded variation), we represent σ as a difference between two nonnegative quasi-volumes p and q, defined on the same dense set of blocks $Q(\sigma)$, as in Sec. 4.7.1. Then, according to Sec. 5.3, the integral $\tilde{I}_\sigma$ is defined as the difference between the nonnegative integrals $\tilde{I}_p$ and $\tilde{I}_q$. It follows that

$$\begin{aligned}\tilde{\sigma}(B) = \tilde{I}_\sigma \chi_B &= \lim_{m\to\infty} I_\sigma f_m = \lim_{m\to\infty} I_p f_m - \lim_{m\to\infty} I_q f_m \\ &= \tilde{p}(B) - \tilde{q}(B) = \lim_{m\to\infty} p(B_m) - \lim_{m\to\infty} q(B_m) = \lim_{m\to\infty} \sigma(B_m),\end{aligned}$$

i.e., formula (12) continues to hold.

Remark 1. It should be emphasized that since σ is of bounded variation, so is $\tilde{\sigma}$, and

$$\int_{\mathbf{B}} f(x)\sigma(dx) = \int_{\mathbf{B}} f(x)\tilde{\sigma}(dx)$$

for every function $f(x)$ continuous in $\mathbf{B}$ (by the very construction of $\tilde{\sigma}$).

Remark 2. In the case $n = 1$, the above results take the following form: Given any generating function of bounded variation defined on a dense set $E \subset [a, b]$ (containing a and b), the right-hand limit

$$\tilde{F}(x) = \lim_{\substack{x+h\in E \\ h\searrow 0}} F(x + h)$$

exists for *every* point $x \in (a, b]$. This is hardly surprising, since, according to Jordan's theorem (p. 85), $F(x)$ can be represented as the difference between two nondecreasing functions, for which the existence of $\tilde{F}(x)$ is obvious, even at the point $x = a$. Moreover, the function $\tilde{F}(x)$ is itself of bounded variation, and the relation

$$\int_a^b f(x)\, dF(x) = \int_a^b f(x)\, d\tilde{F}(x)$$

holds for every function $f(x)$ continuous in $[a, b]$.

5.6. Continuous Quasi-Volumes

A quasi-volume σ (in general, signed), defined on *all* blocks $B \subset \mathbf{B}$, is said to be *continuous* (more exactly, *upper continuous*) if it satisfies the condition

$$\sigma(B) = \lim_{m\to\infty} \sigma(B_m) \tag{13}$$

for every block $B \subset \mathbf{B}$ and every sequence $B_m \searrow B$. In particular, the quasi-volume $\tilde{\sigma}(B)$ defined by formula (6) is continuous. In fact, as already noted, $\tilde{\sigma}(B)$ is defined for all blocks $B \subset \mathbf{B}$, and moreover, substituting $\sigma = \tilde{\sigma}$ into (12) and taking account of (8), we find that (13) holds with $\sigma = \tilde{\sigma}$.

Another important notion is that of continuity on the empty set. A quasi-volume σ (defined on all blocks $B \subset \mathbf{B}$) is said to be *continuous on the empty set* if

$$\lim_{m\to\infty} \sigma(B_m) = 0$$

for every sequence $B_1 \supset B_2 \supset \cdots$ with an empty intersection

$$\bigcap_{m=1}^{\infty} B_m = \varnothing .$$

LEMMA. *If a signed quasi-volume σ of bounded variation is continuous on the empty set, then so are its positive, negative and total variations.*

Proof. Let $\sigma = p - q$, where p and q are the positive and negative variations of σ, and suppose the intersection of the sequence of blocks $B_1 \supset B_2 \supset \cdots$ is empty. Then, as we now show,

$$\lim_{m\to\infty} p(B_m) = 0. \tag{14}$$

Suppose (14) is not true. Then

$$p(B_m) > c$$

for some $c > 0$ and all m. Choose a subsequence B_{m_k} of the sequence B_m according to the following inductive rule, starting from $B_{m_1} = B_1$: Given B_{m_k}, let $B_{m_k}^{(1)}, \ldots, B_{m_k}^{(r_k)}$ be a set of disjoint subblocks of B_{m_k} such that

$$\sum_{j=1}^{r_k} \sigma(B_{m_k}^{(j)}) > \frac{2c}{3} \tag{15}$$

[recall the definition of $p(B)$ on p. 81]. Since the sequence of blocks B_m has an empty intersection, the same is true of every sequence $B_{m_k}^{(j)}B_m$ (for fixed j and k). But then, since $\sigma(B)$ is continuous on the empty set,

$$\sum_{j=1}^{r_k} \sigma(B_{m_k}^{(j)}B_m) < \frac{c}{3}$$

for sufficiently large $m = m_{k+1}$, and hence

$$\sum_{j=1}^{r_k} \sigma(B_{m_k}^{(j)} - B_{m_k}^{(j)}B_{m_{k+1}}) > \frac{c}{3},$$

where $\sigma(B_{m_k}^{(j)} - B_{m_k}^{(j)}B_{m_{k+1}})$ is shorthand for the sum of the quasi-volumes of any disjoint set of blocks with union $B_{m_k}^{(j)} - B_{m_k}^{(j)}B_{m_{k+1}}$ (such a set exists by Property 2, p. 62). This determines the next block $B_{m_{k+1}}$ in the subsequence, which in turn leads to a new set of blocks $B_{m_{k+1}}^{(j)}$ satisfying (15) with k replaced by $k + 1$. By construction, none of the blocks whose union is $B_{m_k}^{(j)} - B_{m_k}^{(j)}B_{m_{k+1}}$ intersects any of the blocks whose union is $B_{m_l}^{(j)} - B_{m_l}^{(j)}B_{m_{l+1}}$ if $k \neq l$. It follows that

$$\sum_{k=1}^{N}\sum_{j=1}^{r_k} \sigma(B_{m_k}^{(j)} - B_{m_k}^{(j)}B_{m_{k+1}}) > \frac{Nc}{3},$$

i.e., the left-hand side can be made as large as we please by choosing N large enough. But this contradicts the assumption that the quasi-volume σ is of bounded variation, and therefore (14) must hold. Thus, if σ is continuous on the empty set, so is p, and hence so are $q = p - \sigma$ and $v = p + q$, as asserted.

The relation between (upper) continuity and continuity on the empty set is revealed by

THEOREM 2. *A quasi-volume σ of bounded variation is continuous if and only if it is continuous on the empty set.*

Proof. If σ is continuous, then

$$\sigma(B) = \tilde{\sigma}(B) = \tilde{I}_\sigma \chi_B. \tag{16}$$

Given any sequence of blocks $B_1 \supset B_2 \supset \cdots$ with an empty intersection, we set $B = B_m$ in (16). Then, noting that $\chi_{B_m} \searrow 0$ and applying Levi's theorem, we find that $\sigma(B_m) \to 0$, i.e., σ is continuous on the empty set.

To prove the converse, we first assume that σ is nonnegative. Let $B_m \searrow B$ and write

$$B = \{x\colon \alpha_1 < x_1 \leqslant \beta_1, \ldots, \alpha_n < x_n \leqslant \beta_n\},$$

$$B_m = \{x\colon \alpha_1^{(m)} < x_1 \leqslant \beta_1^{(m)}, \ldots, \alpha_n^{(m)} < x_n \leqslant \beta_n^{(m)}\},$$

$$\underline{B}_m^{(k)} = \{x\colon \alpha_k < x_k \leqslant \alpha_k^{(m)}\}, \qquad \bar{B}_m^{(k)} = \{x\colon \beta_k < x_k \leqslant \beta_k^{(m)}\}$$

$$(k = 1, \ldots, n).$$

Then, clearly

$$\underline{B}_1^{(k)} \supset \underline{B}_2^{(k)} \supset \cdots, \qquad \bigcap_{m=1}^{\infty} \underline{B}_m^{(k)} = \varnothing,$$

$$\bar{B}_1^{(k)} \supset \bar{B}_2^{(k)} \supset \cdots, \qquad \bigcap_{m=1}^{\infty} \bar{B}_m^{(k)} = \varnothing,$$

and hence, by hypothesis,

$$\lim_{m\to\infty} \sigma(\underline{B}_m^{(k)}) = \lim_{m\to\infty} \sigma(\bar{B}_m^{(k)}) = 0.$$

Moreover, it is easy to see that

$$B \subset B_m \cup \underline{B}_m^{(1)} \cup \cdots \cup \underline{B}_m^{(n)}, \qquad B_m \subset B \cup \bar{B}_m^{(1)} \cup \cdots \cup \bar{B}_m^{(n)},$$

and hence

$$\sigma(B) \leqslant \sigma(B_m) + \sum_{k=1}^{n} \sigma(\underline{B}_m^{(k)}), \tag{17}$$

$$\sigma(B_m) \leqslant \sigma(B) + \sum_{k=1}^{n} \sigma(\bar{B}_m^{(k)}). \tag{18}$$

But (17) and (18) imply

$$\sigma(B) \leqslant \lim_{m\to\infty} \sigma(B_m)$$

and

$$\lim_{m\to\infty} \sigma(B_m) \leqslant \sigma(B),$$

respectively, which in turn imply

$$\sigma(B) = \lim_{m\to\infty} \sigma(B_m),$$

i.e., the quasi-volume σ is continuous.

Finally, we remove the restriction that σ be nonnegative. Suppose σ is signed and continuous on the empty set, with representation $\sigma = p - q$ in terms of the positive and negative variations p and q. By the lemma, p and q are themselves continuous on the empty set. But then p and q are continuous, by the argument just given, and hence so is their difference $\sigma = p - q$. This completes the proof.

Remark 1. Clearly, the lemma remains valid if we omit the phrase "on the empty set."

Remark 2. In the case $n = 1$, the above results take the following form: If the quasi-length σ is continuous, then its generating function $F(x) = \sigma[a, x]$ is defined for every $x \in [a, b]$ (recall that $F(a) = 0$ by definition), and moreover

$$F(x_0) = \sigma[a, x_0] = \lim_{x \searrow x_0} \sigma[a, x] = \lim_{x \searrow x_0} F(x) = F(x_0 + 0) \tag{19}$$

if $x_0 > a$, i.e., $F(x)$ is continuous from the right in $(a, b]$. Conversely, if $F(x)$ is defined in $[a, b]$ and continuous from the right in $(a, b]$, then (19) shows that σ is (upper) continuous in every interval $[a, x_0]$, $x_0 > a$. But σ is also continuous in every other interval $(\alpha_0, \beta_0] \subset [a, b]$, since $\alpha \searrow \alpha_0$, $\beta \searrow \beta_0$ implies

$$\sigma(\alpha, \beta] = F(\beta) - F(\alpha) \to F(\beta_0) - F(\alpha_0) = \sigma(\alpha_0, \beta_0].$$

As for the exceptional point $x = a$ ($[a, a]$ is not a block!), although $F(a + 0)$ always exists, there is no reason for it to coincide with $F(a) = 0$.

Next, as promised in the remark on p. 84, we establish the connection between the canonical representation of functionals given in Sec. 2.11 and the canonical representation of quasi-volumes given in Sec. 4.7:

THEOREM 3. *Let σ be a continuous signed quasi-volume of bounded variation, with canonical representation $\sigma = p - q$, and let I_σ be the corresponding functional on the space $C(\mathbf{B})$,[3] with canonical representation $I_\sigma = J - N$. Then the two representations are consistent, in the sense that*

$$J = I_p, \qquad N = I_q. \tag{20}$$

Proof. According to Theorem 1, the functionals J and N are Riemann-Stieltjes integrals with respect to nonnegative quasi-volumes p_1 and q_1, which are automatically continuous. Therefore the canonical representation of I_σ can be written in the form

$$I_\sigma = I_{p_1} - I_{q_1}. \tag{21}$$

On the other hand, it is obvious that

$$I_\sigma = I_p - I_q.$$

Therefore

$$I_p f \geqslant I_{p_1} f$$

for every $f \in C(\mathbf{B})$, because of the basic minimal property of the canonical representation (21). But this implies

$$p(B) \geqslant p_1(B), \tag{22}$$

since, according to Remark 1, p. 102, the quasi-volume p is continuous. Moreover, according to (21) again, we have $\sigma(B) = p_1(B) - q_1(B)$, and hence

$$p_1(B) \geqslant p(B), \tag{23}$$

by the basic minimal property of the canonical representation $\sigma = p - q$. Comparing (22) and (23), we obtain $p(B) = p_1(B)$ and hence $q(B) = q_1(B)$. Since $J = I_{p_1}$, $N = I_{q_1}$, this implies (20), as required.

5.7. Equivalent Quasi-Volumes

As shown in Sec. 4.4, starting from any quasi-volume σ defined on a dense set Q of blocks $B \subset \mathbf{B}$, we can construct the Riemann-Stieltjes integral

$$I_\sigma f = \int_{\mathbf{B}} f(x)\sigma(dx)$$

[3] I.e., the Riemann-Stieltjes integral with respect to the quasi-volume σ.

of any function $f(x)$ continuous in $\mathbf{B}$. In so doing, we do not exclude the possibility that different quasi-volumes σ_1 and σ_2 may lead to identical values of the integrals $I_{\sigma_1} f$ and $I_{\sigma_2} f$, and hence (for nonnegative quasi-volumes) to identical spaces $\tilde{L}_{\sigma_1}$ and $\tilde{L}_{\sigma_2}$ of Lebesgue-Stieltjes integrable functions. As in Sec. 4.4.4, such quasi-volumes are called *equivalent*. For example, given a quasi-volume σ of bounded variation, we can use Theorem 1 to construct an equivalent *continuous* quasi-volume σ. Moreover, if two equivalent quasi-volumes are continuous, they must coincide. In fact, if $B_m \searrow B$, then, according to Secs. 5.5 and 5.6,

$$\sigma_1(B) = \lim_{m\to\infty} \sigma_1(B_m) = \tilde{\sigma}_1(B), \qquad \sigma_2(B) = \lim_{m\to\infty} \sigma_2(B_m) = \tilde{\sigma}_2(B_m),$$

where $\tilde{\sigma}_1$ and $\tilde{\sigma}_2$ coincide because of the equivalence of σ_1 and σ_2. In other words, the class of all quasi-volumes equivalent to a given quasi-volume contains a unique continuous quasi-volume.

It follows at once from the above considerations that two quasi-volumes σ_1 and σ_2 are equivalent if and only if $B'_m \searrow B$, $B''_m \searrow B$, $B'_m \in Q(\sigma_1)$, $B''_m \in Q(\sigma_2)$ implies

$$\lim_{m\to\infty} \sigma_1(B'_m) = \lim_{m\to\infty} \sigma_2(B''_m).$$

For nonnegative quasi-volumes, we can say even more:

THEOREM 4. *Two nonnegative quasi-volumes σ_1 and σ_2 are equivalent if and only if $B' \Subset B''$ implies*

$$\begin{aligned} \sigma_1(B') &\leqslant \sigma_2(B'') \qquad \textit{if} \quad B' \in Q_1,\ B'' \in Q_2, \\ \sigma_2(B') &\leqslant \sigma_1(B'') \qquad \textit{if} \quad B' \in Q_2,\ B'' \in Q_1. \end{aligned} \tag{24}$$

Proof. Suppose σ_1 and σ_2 are equivalent. Choosing blocks $B' \in Q_1$, $B'' \in Q_2$, $B' \Subset B''$, let $f(x)$ be a continuous function, taking values between 0 and 1, such that

$$f(x) = \begin{cases} 1 & \text{for} \quad x \in B', \\ 0 & \text{for} \quad x \notin B''. \end{cases}$$

Then, obviously,

$$\sigma_1(B') \leqslant \int_{\mathbf{B}} f(x)\sigma_1(dx) = \int_{\mathbf{B}} f(x)\sigma_2(dx) \leqslant \sigma_2(B''),$$

as required.

Conversely, suppose (24) holds for two quasi-volumes σ_1 and σ_2. Then, starting from the Riemann-Stieltjes integrals, we can use Theorem 1 to construct the corresponding quasi-volumes $\tilde{\sigma}_1$ and $\tilde{\sigma}_2$. According to Sec. 5.5, given any block $B \subset \mathbf{B}$,

$$\tilde{\sigma}_1(B) = \lim_{m\to\infty} \sigma_1(B_m^{(1)}), \qquad \tilde{\sigma}_2(B) = \lim_{m\to\infty} \sigma_2(B_m^{(2)}),$$

where $B_m^{(1)} \in Q_1$ and $B_m^{(2)} \in Q_2$ are any two sequences of blocks converging downward to the block B. Given a sequence $B_m^{(1)} \searrow B$, we can always find another sequence $B_m^{(2)} \searrow B$ such that

$$B_m^{(2)} \Subset B_m^{(1)} \qquad (m = 1, 2, \ldots).$$

Then, by hypothesis,

$$\sigma_2(B_m^{(2)}) \leqslant \sigma_1(B_m^{(1)}),$$

and hence

$$\tilde{\sigma}_2(B) = \lim_{m\to\infty} \sigma_2(B_m^{(2)}) \leqslant \lim_{m\to\infty} \sigma_1(B_m^{(1)}) = \tilde{\sigma}_1(B). \tag{25}$$

On the other hand, we could just as well have found a sequence $B_m^{(2)} \searrow B$ satisfying the opposite inclusion relation

$$B_m^{(2)} \Supset B_m^{(1)},$$

and then

$$\tilde{\sigma}_2(B) = \lim_{m\to\infty} \sigma_2(B_m^{(2)}) \geqslant \lim_{m\to\infty} \sigma_1(B_m^{(1)}) = \tilde{\sigma}_1(B), \tag{26}$$

since $\tilde{\sigma}_2(B)$ does not depend on the choice of the sequence $B_m^{(2)}$. Together, (25) and (26) imply $\tilde{\sigma}_1(B) = \tilde{\sigma}_2(B)$ for any block B. But since σ_1 is equivalent to $\tilde{\sigma}_1$ and σ_2 is equivalent to $\tilde{\sigma}_2$, it follows that σ_1 is equivalent to σ_2, as asserted.

Remark. In the case $n = 1$, the above results take the following form: Let $\sigma_1(\alpha, \beta]$ and $\sigma_2(\alpha, \beta]$ be two quasi-lengths of bounded variation, with corresponding generating functions $F(x)$ and $G(x)$, defined on dense sets E_1 and E_2 (in $[a, b]$). Then σ_1 and σ_2 are equivalent if and only if $\tilde{F}(x) = \tilde{G}(x)$ for every $x \in [a, b]$, where $\tilde{F}$ and $\tilde{G}$ are defined as in Remark 2, p. 99. Moreover, if σ_1 and σ_2 are nonnegative, they are equivalent if and only if $x' < x''$ implies

$$F_1(x') \leqslant F_2(x'') \qquad \text{if} \quad x' \in E_1,\ x'' \in E_2,$$
$$F_2(x') \leqslant F_1(x'') \qquad \text{if} \quad x' \in E_2,\ x'' \in E_1.$$

5.8. Construction of the Lebesgue-Stieltjes Integral with Step Functions as Elementary Functions

Suppose we want to construct the Lebesgue-Stieltjes integral with respect to a given quasi-volume σ, choosing the elementary functions to be step functions (as in Sec. 3.1, for the ordinary Lebesgue integral). Then the natural choice of the elementary integral is

$$I_\sigma h = \sum_{j=1}^{m} h_j \sigma(B_j), \qquad B_j = \{x : h(x) = h_j\}, \tag{27}$$

which differs from formula (1), p. 50 only by having σ in place of s (as usual, $\{B_j\}$ is a partition of the basic block **B**). We begin by assuming that σ is nonnegative, removing this restriction at the end of the section. Before we can apply the general Daniell scheme of Chap. 2 (with H the family of step functions defined on the set $X = \mathbf{B}$), it must be verified that I_σ satisfies all the axioms for an elementary integral given in Sec. 2.1. Axioms 1 and 2 are obvious, but we must be careful about Axiom 3, which asserts that if $h_m(x) \searrow 0$, then $I_\sigma h_m \to 0$. In fact, it turns out that Axiom 3 is not satisfied unless an extra condition is imposed on the quasi-volume σ:

THEOREM 5. *The proposed elementary integral I_σ satisfies Axiom 3 if and only if the quasi-volume σ is continuous.*

Proof. Let B be any block and $B_m \searrow B$ any sequence of blocks converging downward to B. If I_σ satisfies Axiom 3 (besides Axioms 1 and 2), the Daniell construction will lead to an integral satisfying Lebesgue's theorem. But then, since the sequence of characteristic functions $\chi_{B_m}(x)$ obviously converges (everywhere) to the characteristic function $\chi_B(x)$, we have

$$\sigma(B) = I_\sigma \chi_B = \lim_{m\to\infty} I_\sigma \chi_{B_m} = \lim_{m\to\infty} \sigma(B_m),$$

and hence σ is continuous.

Conversely, suppose σ is continuous. It would be difficult to verify Axiom 3 directly, as in Sec. 1.5, since the functions $h_m(x)$ and the quasi-volume σ may share "sheets of discontinuity."[4] However, an indirect proof, involving the integral $\tilde{I}_\sigma$ and the space $\tilde{L}_\sigma$ constructed in Sec. 5.1, can easily be given. In fact, let

$$h_m(x) = \sum_{j=1}^{r_m} h_j^{(m)} \chi_{B_j^{(m)}}(x)$$

be a nonincreasing sequence of step functions converging to zero. Then, since $\tilde{L}_\sigma$ contains all step functions (cf. p. 95), and since σ is continuous

---

[4] A sheet S is said to be a *sheet of discontinuity* of the (nonnegative) quasi-volume σ if there exists a sequence of blocks $B_1 \supset B_2 \supset \cdots$ whose intersection

$$B = \bigcap_{m=1}^{\infty} B_m$$

is contained in S, such that $\sigma(B) > 0$. There is nothing to prevent an (upper) continuous quasi-volume σ from having sheets of discontinuity. For example, suppose σ is the continuous quasi-length with generating function

$$F(x) = \begin{cases} 0 & \text{for } 0 \leqslant x < 1, \\ 1 & \text{for } 1 \leqslant x \leqslant 2. \end{cases}$$

Then the point $x = 1$ is a sheet of discontinuity (i.e., a discontinuity point) of σ.

($\sigma = \tilde{\sigma}$), we have

$$I_\sigma h_m = \sum_{j=1}^{r_m} h_j^{(m)} \sigma(B_j^{(m)}) = \sum_{j=1}^{r_m} h_j^{(m)} \tilde{\sigma}(B_j^{(m)}) = \tilde{I}_\sigma h_m.$$

But Levi's theorem holds in the space $\tilde{L}_\sigma$, and hence $\tilde{I}_\sigma h_m \to 0$, i.e., $I_\sigma h_m \to 0$, as required.

Thus all the prerequisites for constructing a theory of the integral, based on step functions as elementary functions, with the integral (27) as elementary integral, are satisfied. Let $L_\sigma = L_\sigma(\mathbf{B})$ denote the corresponding space of σ-summable functions, equipped with a Lebesgue-Stieltjes integral $I_\sigma f$. Then, as we now show, this construction of the Lebesgue-Stieltjes integral agrees with that of Sec. 5.1, based on continuous functions as elementary functions with the Riemann-Stieltjes integral as elementary integral, leading to a space $\tilde{L}_\sigma$ of σ-summable functions equipped with a Lebesgue-Stieltjes integral $I_\sigma f$. More precisely, we prove the following

THEOREM 6. *The two constructions of the Lebesgue-Stieltjes integral are equivalent, i.e.,* $\tilde{L}_\sigma = L_\sigma$ *and* $\tilde{I}_\sigma f = I_\sigma f$.

Proof. As might be expected, the proof is word for word the same as that of the theorem on p. 54, if we make the following substitutions:

$$L \to L_\sigma, \quad \tilde{L} \to \tilde{L}_\sigma, \quad I \to I_\sigma, \quad \tilde{I} \to \tilde{I}_\sigma, \quad s(B_j) \to \sigma(B_j), \quad dx \to \sigma(dx).$$

Finally, as promised, we remove the restriction that σ be nonnegative. Let σ be any continuous *signed* quasi-volume of bounded variation, with canonical representation $\sigma = p - q$ and total variation $v = p + q$. Then, according to Sec. 5.3, the spaces L_σ and $\tilde{L}_\sigma$ figuring in Theorem 6 should be replaced by L_v and $\tilde{L}_v$, while the integrals I_σ and $\tilde{I}_\sigma$ take the form

$$I_\sigma = I_p - I_q, \qquad \tilde{I}_\sigma = \tilde{I}_p - \tilde{I}_q.$$

(Note that Theorem 3 guarantees the same integral, whether we follow the procedure of Sec. 2.11.2 or that of Sec. 5.3.) But, according to Theorem 6, $\tilde{L}_p = L_p, \tilde{L}_q = L_q, \tilde{L}_v = L_v$ and $\tilde{I}_p f = I_p f, \tilde{I}_q f = I_q f$. It follows that $\tilde{I}_\sigma f = I_\sigma f$, as required.

PROBLEMS

1. Let $s(B)$ be ordinary volume, and let $\sigma(B)$ be the quasi-volume defined by

$$\sigma(B) = \begin{cases} 1 & \text{for } x_0 \in B, \\ 0 & \text{for } x_0 \notin B, \end{cases}$$

where x_0 is a fixed point and $\mathbf{B}$ is finite. Find a function summable with respect to $\sigma(B)$, but not with respect to $s(B)$.

Ans. For example,

$$\varphi(x) = \begin{cases} 0 & \text{for} \quad x = x_0, \\ \dfrac{1}{|x - x_0|^n} & \text{for} \quad x \neq x_0. \end{cases}$$

2. Let $s(B)$ and $\sigma(B)$ be the same as in the preceding problem. Find a function summable with respect to $s(B)$, but not with respect to $\sigma(B)$.

Ans. For example,

$$\varphi(x) = \begin{cases} \infty & \text{for} \quad x = x_0, \\ 0 & \text{for} \quad x \neq x_0. \end{cases}$$

3. Let $\sigma(\alpha, \beta]$ be the quasi-length defined on the closed interval $[0, 1]$ by the formula

$$\sigma(\alpha, \beta] = \int_\alpha^\beta \ln \frac{1}{x}\, dx.$$

Find a function $\varphi(x)$ which is Lebesgue-integrable (in the ordinary sense) but not σ-summable, and a function $\psi(x)$ which is σ-summable but not Lebesgue-integrable.

Ans. For example,

$$\varphi(x) = \frac{1}{x \ln^2 x}, \qquad \psi(x) = \frac{1}{1 - x}.$$

4. Evaluate the Stieltjes integral

$$I = \int_0^3 \varphi(x)\, dF(x),$$

where

$$\varphi(x) = \begin{cases} 1 & \text{for} \quad 0 \leqslant x \leqslant 1, \\ 0 & \text{for} \quad 1 < x \leqslant 3, \end{cases} \qquad F(x) = \begin{cases} -1 & \text{for} \quad 0 \leqslant x \leqslant 1, \\ 2 & \text{for} \quad 1 < x \leqslant 2, \\ -2 & \text{for} \quad 2 < x \leqslant 3. \end{cases}$$

Ans. $I = 1$.

5. Evaluate the Stieltjes integral

$$I = \int_0^1 \varphi(x)\, dF(x),$$

where

$$\varphi(x) = \begin{cases} 1 & \text{for} \quad 0 \leqslant x \leqslant \frac{1}{2}, \\ 0 & \text{for} \quad \frac{1}{2} < x \leqslant 1, \end{cases} \qquad F(x) = \begin{cases} 0 & \text{for} \quad 0 \leqslant x \leqslant \frac{1}{2}, \\ 1 & \text{for} \quad \frac{1}{2} < x \leqslant 1. \end{cases}$$

Ans. $I = 1$(!).

6. Suppose one tries to construct a theory of Stieltjes integration in which the basic block **B** and *all* its subblocks are half-open sets of the form

$$\{x\colon \alpha_1 < x_1 \leqslant \beta_1, \ldots, \alpha_n < x_n \leqslant \beta_n\}.$$

Show that the analogue of Theorem 1, p. 95 fails to hold for the space $C(\mathbf{B})$ of all functions uniformly continuous in $\mathbf{B}$.

Hint. Let $\mathbf{B} = (0, 1]$, $\sigma(0, \beta] \equiv 1$, $\sigma(\alpha, \beta] = 0$ if $\alpha > 0$, and consider the sequence

$$f_m(x) = \begin{cases} 1 & \text{for } 0 < x < \dfrac{1}{m}, \\ m\left(\dfrac{2}{m} - x\right) & \text{for } \dfrac{1}{m} \leqslant x < \dfrac{2}{m}, \\ 0 & \text{for } \dfrac{2}{m} < x \leqslant 1 \end{cases}$$

of functions uniformly continuous in $\mathbf{B}$. Then $f_m(x) \searrow 0$ for all $x \in \mathbf{B}$, but $I_\sigma f_m = 1$ for all m.

7. Let $\sigma(\alpha, \beta]$ be the quasi-length defined on the closed interval $[0, 2]$ by the formula

$$\sigma(\alpha, \beta] = \begin{cases} 1 & \text{if } \; x_0 = 1 \text{ is an } \textit{interior} \text{ point of } (\alpha, \beta], \\ 0 & \text{otherwise.} \end{cases}$$

Show that if $B = [0, 1]$, then $\sigma(B) = 0$ but $\tilde{\sigma}(B) = 1$.

8. There is a proof of Theorem 1, p. 95 based on the Hahn-Banach theorem, according to which a linear functional can be extended from a given normed linear space R to any larger normed linear space $R' \supset R$. In fact, the Hahn-Banach theorem is used to extend the given linear functional If from its original domain $C(\mathbf{B})$ to the space $S(\mathbf{B})$ of all bounded functions $f(x)$ with the same norm

$$\sup_{x \in \mathbf{B}} |f(x)| .$$

The values of the extended functional on characteristic functions of blocks defines a quasi-volume $\sigma(B)$ of bounded variation, which is in general not continuous. The quasi-volume $\sigma(B)$ can then be used to define a Riemann-Stieltjes integral which coincides with If on continuous functions. Going from $\sigma(B)$ to $\tilde{\sigma}(B)$ as in Sec. 5.5 makes the quasi-volume continuous without changing Riemann-Stieltjes integrals of continuous functions. Carry out this construction (due to L. A. Lusternik) in detail.

Part 3

MEASURE

6

MEASURABLE SETS AND GENERAL MEASURE THEORY

In constructing a theory of the integral, our first step was to define the volume (or quasi-volume) of certain "elementary figures," namely blocks. We now use our fully developed theory of the integral to construct the volume (or quasi-volume) of "nonelementary figures," i.e., sets of a more or less general nature. As might be expected, the "volume" of a figure M should be defined as the integral of its characteristic function $\chi_M(x)$, equal to 1 on M and 0 outside M. This approach leads to certain difficulties, which can be circumvented in a way that becomes more transparent if we adopt an abstract point of view.

6.1. More on Measurable Functions

As in Sec. 2.1 we start from a family H of elementary functions, defined on an abstract set X, which satisfy Axioms a and b on p. 23. One of our main concerns in this chapter will be the class of measurable functions, already encountered incidentally in Part 1. According to Sec. 2.8.1, a function $f(x)$, defined on a set X, is called *measurable* if it is finite almost everywhere and can be represented on a set of full measure as the limit of a convergent sequence of elementary functions. In the first instance, this definition implies that any function differing from a measurable function only on a set of measure zero is itself measurable. The following are some further consequences of the definition of measurability:

1) If f and g are measurable, so that $f = \lim h_n$ on a set of full measure E and $g = \lim k_n$ on a set of full measure F, then, given any real numbers

α and β, we have $\alpha f + \beta g = \lim(\alpha h_n + \beta k_n)$ on a set of full measure EF (the intersection of E and F), and hence $\alpha f + \beta g$ is also measurable.

2) If $f = \lim h_n$ is measurable, so is $|f| = \lim |h_n|$. It follows that if f is measurable, so are f^+ and f^-, and hence, if f and g are measurable, so are

$$\max(f, g) = (f - g)^+ + g^+, \qquad \min(f, g) = -\max(-f, -g).$$

3) Every function $f \in L^+$ is measurable, being the limit of a sequence of elementary functions (in fact, a nondecreasing sequence). Moreover, every summable function $\varphi \in L$ is measurable, since $\varphi = f - g$ where $f, g \in L^+$.

THEOREM 1. *If f_n is a nondecreasing sequence of summable functions converging almost everywhere to a finite limit f, then f is measurable.*[1]

Proof. The proof bears a resemblance to those of Theorems 1 and 2 of Chap. 2. First suppose that $f_n \in L^+$, and let h_{nk} $(k = 1, 2, \ldots)$ be a sequence of elementary functions such that $h_{nk} \nearrow f_n$ as $k \to \infty$. If

$$h_n = \max(h_{1n}, \ldots, h_{nn}),$$

then $h_n(x)$ is a (nondecreasing) sequence of elementary functions, and hence has a finite or infinite limit $f^*(x)$ for every $x \in X$. For $n \geqslant k$, we have

$$h_{kn} \leqslant h_n \leqslant f_n.$$

Taking the limit of this inequality as $n \to \infty$, we obtain

$$f_k \leqslant f^* \leqslant f,$$

which shows that f^*, like f, is finite almost everywhere. Therefore f^* is measurable. But $f^* = f$ almost everywhere, since $f_k \nearrow f$, and hence f is measurable.

Now let the f_n be arbitrary summable functions. We have

$$f = \lim_{n \to \infty} f_n = \sum_{n=0}^{\infty} \varphi_n,$$

where

$$\varphi_0 = f_1, \varphi_1 = f_2 - f_1, \ldots, \varphi_n = f_{n+1} - f_n, \ldots$$

are nonnegative summable functions. According to the final observation of Sec. 2.5, we can set $\varphi_n = g_n - g_n^*$, where the functions g_n and g_n^* are nonnegative and belong to L^+, and moreover

$$I g_n^* \leqslant \frac{1}{2^n}.$$

[1] Although, in general, not summable.

The sum of the series

$$\sum_{n=0}^{\infty} g_n^*$$

is a function $g^* \in L^+$, while the series

$$\sum_{n=0}^{\infty} g_n = \sum_{n=0}^{\infty} \varphi_n + \sum_{n=0}^{\infty} g_n^*$$

converges almost everywhere to a finite limit g, which, according to what was just proved, is a measurable function. Therefore

$$f = \sum_{n=0}^{\infty} \varphi_n = \sum_{n=0}^{\infty} g_n - \sum_{n=0}^{\infty} g_n^* = g - g^*$$

is measurable, as asserted.

The example

$$\varphi(x) = \frac{1}{x}, \qquad (0 \leqslant x \leqslant 1)$$

shows that there are measurable functions which are not summable. However, as remarked in Sec. 2.8.1, if a measurable function $\varphi(x)$ satisfies the inequality

$$|\varphi(x)| \leqslant \varphi_0(x),$$

where $\varphi_0(x)$ is a nonnegative summable function, then $\varphi(x)$ is summable. Using this fact, we can prove the converse of Theorem 1 for nonnegative measurable functions, i.e., *every nonnegative measurable function is the limit of a nondecreasing sequence of summable functions.* Indeed, if

$$f(x) = \lim_{n\to\infty} h_n(x) \geqslant 0,$$

and if $h_n(x) \geqslant 0$ (as can be assumed without loss of generality), then

$$f(x) = \lim_{n\to\infty} f_n(x) \equiv \lim_{n\to\infty} \min \{f(x), \max [h_1(x), \ldots, h_n(x)]\}.$$

The function $f_n(x)$ is summable, since it is measurable, nonnegative and no greater than the summable function $\max \{h_1(x), \ldots, h_n(x)\}$. Moreover, it is obvious that $f_n(x) \nearrow f(x)$ as $n \to \infty$.

THEOREM 2. *If f_n is a sequence of measurable functions converging almost everywhere to a function f, then f is measurable.*

Proof. There is no loss of generality in restricting f and f_n to be nonnegative (otherwise, consider $f^+ = \lim f_n^+$ and $f^- = \lim f_n^-$ separately). The measurable function $f_n(x)$ is the limit as $p \to \infty$ of a sequence of elementary functions $h_p^{(n)}(x)$, which can be assumed to be nonnegative with positive integrals. Consider the function

$$\varphi_0(x) = \sum_{n,p=1}^{\infty} c_p^{(n)} \frac{h_p^{(n)}(x)}{Ih_p^{(n)}}, \tag{1}$$

where the (positive) coefficients $c_p^{(n)}$ are such that the series

$$\sum_{n,p=1}^{\infty} c_p^{(n)}$$

converges. Then, by Levi's theorem, $\varphi_0(x)$ is summable, since, by construction, the series of integrals of the separate terms in the right-hand side of (1) converges. Obviously, $\varphi_0(x) > 0$ wherever $f_0(x) > 0$. It follows that $f(x)$ can also be represented as the limit of the nondecreasing sequence

$$g_n(x) = \min \{f(x), n\varphi_0(x)\}.$$

Because of Theorem 1, we need only verify that measurability of $f_n(x)$ implies summability of $g_n(x)$. But clearly

$$g_n(x) = \min \{f(x), n\varphi_0(x)\} = \lim_{m\to\infty} \min \{f_m(x), n\varphi_0(x)\},$$

where the functions $\min \{f_m(x), n\varphi_0(x)\}$, $m = 1, 2, \ldots$ are measurable and bounded by the summable function $n\varphi_0(x)$, and hence themselves summable. It follows from Lebesgue's theorem that their limit function $g_n(x)$ is also summable, and the theorem is proved.

COROLLARY. *Let $f_1(x), f_2(x), \ldots$ be an arbitrary sequence of measurable functions. Then the functions*

$$\inf_n f_n(x) = \lim_{n\to\infty} \inf \{f_1(x), \ldots, f_n(x)\},$$

$$\sup_n f_n(x) = \lim_{n\to\infty} \sup \{f_1(x), \ldots, f_n(x)\}$$

are also measurable, if they are finite almost everywhere, and the same is true of the functions

$$\varliminf_{n\to\infty} f_n(x) = \lim_{n\to\infty} \inf \{f_n(x), f_{n+1}(x), \ldots\},$$

$$\varlimsup_{n\to\infty} f_n(x) = \lim_{n\to\infty} \sup \{f_n(x), f_{n+1}(x), \ldots\}.$$

6.2. Measurable Sets

A set $E \subset X$ is said to be *measurable* if its characteristic function $\chi_E(x)$ (equal to 1 on E and 0 outside E) is measurable. If the function $\chi_E(x)$ is summable as well as measurable, the set E is said to be *summable*, and the number $\mu(E) = I\chi_E$ is called the *measure* of the set E. If a set is measurable but not summable, its measure is taken to be $+\infty$. No measure, finite or infinite, is assigned to a nonmeasurable set.

A measurable subset of a summable set is summable (its characteristic function is summable, being measurable and bounded by a summable function). Any subset of a set of measure zero is measurable and has measure zero (as must be expected!). The empty set is regarded as measurable and summable, and is assigned the measure zero.

The formulas

$$\chi_{E \cup F} = \max(\chi_E, \chi_F),$$
$$\chi_{EF} = \min(\chi_E, \chi_F),$$
$$\chi_{E-F} = \chi_E - \chi_F \qquad (E \supset F)$$

show that *the union, intersection and difference of two measurable sets are measurable.* Similarly, *the union, intersection and difference of two summable sets are summable, and moreover*

$$\mu(E) \leqslant \mu(F) \qquad (E \subset F),$$
$$\mu(E \cup F) \leqslant \mu(E) + \mu(F),$$
$$\mu(E \cup F) = \mu(E) + \mu(F) \qquad (EF = \varnothing),$$
$$\mu(E - F) = \mu(E) - \mu(F) \qquad (E \supset F).$$

6.3. Countable Additivity of Measure

A key proposition of measure theory is

THEOREM 3. *If the sets $E_1, \ldots, E_n, \ldots$ are measurable, then their union*

$$E = \bigcup_{n=1}^{\infty} E_n$$

is measurable. Moreover, measure is countably additive in the sense that if the sets $E_1, \ldots, E_n, \ldots$ are disjoint, then

$$\mu(E) = \mu(E_1) + \cdots + \mu(E_n) + \cdots, \tag{2}$$

where (2) *may reduce to* $\infty = \infty$.

Proof. By hypothesis, each set E_n has a measurable characteristic function $\chi_{E_n}(x)$. Therefore, by the corollary to Theorem 2, the characteristic function of the set E, i.e.,

$$\chi_E(x) = \sup_n \{\chi_{E_1}(x), \ldots, \chi_{E_n}(x), \ldots\} = \lim_{n\to\infty} \sup \{\chi_{E_1}(x), \ldots, \chi_{E_n}(x)\},$$

is also measurable, and hence E is a measurable set, as asserted.

To prove the countable additivity (2), we first note that if some $\mu(E_n)$ is infinite then so is $\mu(E)$, since $E \supset E_n$, and hence (2) reduces to $\infty = \infty$. Therefore we can assume that all the E_n are summable, with $\mu(E_n) = I\chi_{E_n}$. If the E_n are disjoint, then

$$\chi_E(x) = \sum_{n=1}^{\infty} \chi_{E_n}(x).$$

It follows from Levi's theorem that χ_E is summable, with

$$I\chi_E = \sum_{n=1}^{\infty} I\chi_{E_n},$$

if the series

$$\sum_{n=1}^{\infty} I\chi_{E_n} = \sum_{n=1}^{\infty} \mu(E_n) \tag{3}$$

converges. Conversely, if χ_E is summable, then

$$\sum_{n=1}^{N} I\chi_{E_n} \leqslant I\chi_E$$

for any N, and hence the series (3) converges, i.e., χ_E is not summable if (3) diverges. Equation (2) holds in either case, and the theorem is proved.

COROLLARY. *If the sets $E_1, \ldots, E_n, \ldots$ are measurable and $E_1 \subset E_2 \subset \cdots$, then their union*

$$E = \bigcup_{n=1}^{\infty} E_n$$

is measurable, and

$$\mu(E) = \lim_{n\to\infty} \mu(E_n), \tag{4}$$

where (4) may reduce to $\infty = \infty$.

Proof If some $\mu(E_n)$ is infinite, then so are $\mu(E)$ and $\lim \mu(E_n)$ [since $\mu(E_{n+p}) = \infty$ for all p], and hence (4) reduces to $\infty = \infty$. Otherwise, the formula

$$E = E_1 \cup (E_2 - E_1) \cup \cdots$$

represents E as a union of disjoint measurable sets, and therefore, by countable additivity,

$$\mu(E) = \mu(E_1) + \mu(E_2 - E_1) + \cdots = \lim_{n\to\infty} \sum_{k=1}^{n} \mu(E_k - E_{k-1})$$

$$= \lim_{n\to\infty} \mu(E_n) \quad (E_0 = \varnothing),$$

as required.

THEOREM 4. *If the sets $E_1, \ldots, E_n, \ldots$ are measurable, then their intersection*

$$F = \bigcap_{n=1}^{\infty} E_n$$

is measurable. Moreover, if $E_1 \supset E_2 \supset \cdots$ and $\mu(E_1) < \infty$, then

$$\mu(F) = \lim_{n \to \infty} \mu(E_n),$$

where the condition $\mu(E_1) < \infty$ cannot be dropped.[2]

Proof. The first assertion follows at once from Theorem 3, after taking complements relative to E_1. To prove the second assertion, we represent E_1 as a union of disjoint measurable sets by writing

$$E_1 = F \cup (E_1 - E_2) \cup (E_2 - E_3) \cup \cdots,$$

and then use countable additivity.

6.4. Stone's Axioms

In addition to Axioms a and b, p. 23, we shall henceforth impose two further axioms, called *Stone's axioms*, on the family H of elementary functions:

c) If $h(x)$ belongs to H, then so does the function $\min \{h(x), 1\}$, i.e., the function $h(x)$ truncated above the level 1.

d) There exists a sequence of nonnegative functions $h_n(x) \in H$ such that $Ih_n > 0$ and $\sup_n h_n(x) > 0$ for every $x \in X$.

Both axioms hold automatically if H contains the function identically equal to 1, a case which occurs whenever the space X is of finite "volume" $\mu(X) = I(1)$. However, we want the general case to include integration over spaces of infinite volume.

Axiom c also applies to measurable functions, as we see by passing to the limit. Thus if $\varphi = \lim h_n$ is measurable, so is $\min (\varphi, 1) = \lim \min (h_n, 1)$.

Axiom d implies the existence of a summable function $\varphi_0(x)$ which is positive for all $x \in X$. In fact, the series

$$\varphi_0(x) = \sum_{n=1}^{\infty} \frac{1}{n^2} \frac{h_n(x)}{Ih_n},$$

where the $h_n(x)$ are the elementary functions figuring in the axiom, converges to a summable function, by Levi's theorem.

---

[2] See Prob. 2, p. 131.

The presence of the function $\varphi_0(x)$ allows us to deduce some new facts about the class of measurable functions. First of all, the function $f(x) \equiv 1$ is measurable, since

$$1 = \lim_{n \to \infty} \min \{1, n\varphi_0(x)\}$$

and Axiom c (valid, as noted, for all measurable functions) guarantees the measurability of the functions $\min \{1, n\varphi_0(x)\}$. Therefore $f(x) \equiv 1$ is also measurable, by Theorem 2. This implies the measurability of the space X itself, since X has the characteristic function $f(x) \equiv 1$. Then the complement $X - E$ of any measurable set E (relative to the whole space) is measurable, being the difference between two measurable sets. Moreover, if $f(x) \equiv 1$ is measurable, so is any constant function $f(x) \equiv c$. In particular, if φ is measurable and a, b, c are any real numbers, then the following functions are all measurable:

1) $\min (\varphi, c)$, the function $\varphi(x)$ truncated above the level c;
2) $\max (\varphi, c)$, the function $\varphi(x)$ truncated below the level c;
3) $\max \{\min (\varphi, b), a\}$, the function $\varphi(x)$ truncated above the level b and below the level a.

6.5. Characterization of Measurable Functions in Terms of Measure

The relation between measurable functions and measurable sets is revealed further by

THEOREM 5. *An almost-everywhere finite function $\varphi(x)$ is measurable if and only if the set*

$$E(\varphi; c) = \{x: \varphi(x) > c\}$$

is measurable for arbitrary real c.

Proof. If $\varphi(x)$ is measurable, the function

$$\varphi_{\varepsilon c}(x) = \frac{\min (\varphi, c + \varepsilon) - \min (\varphi, c)}{\varepsilon}$$

is measurable for arbitrary c and ε. The function $\varphi_{\varepsilon c}(x)$ equals 0 for $\varphi(x) \leqslant c$ and 1 for $\varphi(x) \geqslant c + \varepsilon$, and takes values between 0 and 1. As $\varepsilon \to 0$, $\varphi_{\varepsilon c}(x)$ approaches a limit equal to 0 for $\varphi(x) \leqslant c$ and 1 for $\varphi(x) > c$. Thus the characteristic functions $\chi_{E(\varphi;\, c)}$ of the set $E(\varphi; c)$ is a limit of measurable functions, which implies the measurability of $\chi_{E(\varphi;\, c)}$ and hence of $E(\varphi; c)$.

Conversely, suppose we know that the set $E(\varphi; c)$ is measurable for arbitrary c. Then the set

$$\{x: c < \varphi(x) \leqslant d\} = E(\varphi; c) - E(\varphi; d)$$

is also measurable, for arbitrary c and d ($c < d$). Given any n, consider the function $\varphi_n(x)$ equal to k/n on the measurable set

$$E_{k,n}(\varphi) = \left\{x: \frac{k}{n} < \varphi(x) \leqslant \frac{k+1}{n}\right\} \qquad (k = 0, \pm 1, \pm 2, \ldots).$$

The function $\varphi_n(x)$ is defined for almost all x, and differs from $\varphi(x)$ by no more than $1/n$. Moreover, $\varphi_n(x)$ can be written in the form

$$\varphi_n(x) = \sum_{k=-\infty}^{\infty} \frac{k}{n} \chi_{E_{k,n}(\varphi)}(x),$$

and hence is measurable. But then $\varphi(x)$ is also measurable, since $\varphi_n(x) \to \varphi(x)$ [uniformly on X] as $n \to \infty$. This completes the proof.

Remark 1. If φ is a nonnegative summable function, then

$$0 \leqslant \min(\varphi, c) \leqslant \varphi$$

for any $c > 0$, and hence $f = \min(\varphi, c)$ is also summable. This implies the summability of the set

$$E(\varphi; c) = \{x: \varphi(x) > c\},$$

already known to be measurable from Theorem 5. In fact, the characteristic function of $E(\varphi; c)$ is summable, since it satisfies the inequality

$$0 \leqslant \chi_{E(\varphi;c)}(x) \leqslant \frac{1}{c} f(x).$$

Remark 2. We have already seen that $f(x) \equiv 1$ is measurable. If $f(x) \equiv 1$ is also summable, then so is the whole space X, and $\mu(X) = I(1)$. In general, $f(x) \equiv 1$ is measurable but not summable, and $\mu(X) = +\infty$. In this case, it is easy to see that X is the union of an increasing sequence of summable sets, i.e., the sets

$$E_n = \left\{x: \varphi_0(x) > \frac{1}{n}\right\},$$

where $\varphi_0(x)$ is the everywhere-positive summable function constructed in Sec. 6.4.

6.6. The Lebesgue Integral as Defined by Lebesgue

We are now in a position to relate the definition of the integral of a summable function φ to the measures of certain summable sets constructed from φ. First suppose φ is nonnegative. Then, given any $\varepsilon > 0$, consider the sets

$$E_{\varepsilon n} = \{x: n\varepsilon < \varphi(x) \leqslant (n+1)\varepsilon\} \qquad (n = 0, 1, 2, \ldots). \tag{5}$$

If $n \geqslant 1$, $E_{\varepsilon n}$ is summable, being the difference between the two summable sets $\{x: \varphi(x) > n\varepsilon\}$ and $\{x: \varphi(x) > (n+1)\varepsilon\}$ (see Remark 1 above). Consider the function $\varphi_\varepsilon(x)$ equal to $n\varepsilon$ on $E_{\varepsilon n}$ $(n = 0, 1, 2, \ldots)$ and 0 elsewhere [i.e., where $\varphi(x)$ itself vanishes]. Then

$$\varphi_\varepsilon(x) = \sum_{n=1}^{\infty} n\varepsilon\chi_{\varepsilon n}(x), \tag{6}$$

where $\chi_{\varepsilon n}(x)$ is the characteristic function of the set $E_{\varepsilon n}$. By Lebesgue's theorem, $\varphi_\varepsilon(x)$ is summable, since $\varphi_\varepsilon(x) \leqslant \varphi(x)$, and

$$I\varphi_\varepsilon(x) = \sum_{n=1}^{\infty} n\varepsilon I\chi_{\varepsilon n} = \sum_{n=1}^{\infty} n\varepsilon\mu(E_{\varepsilon n}). \tag{7}$$

Moreover $0 \leqslant \varphi(x) - \varphi_\varepsilon(x) \leqslant \varepsilon$, and hence $\varphi_\varepsilon(x) \to \varphi(x)$ as $\varepsilon \to 0$. Therefore, applying Lebesgue's theorem again, we find that[3]

$$I\varphi = \lim_{\varepsilon \to 0} I\varphi_\varepsilon = \lim_{\varepsilon \to 0} \sum_{n=1}^{\infty} n\varepsilon\mu\{x: n\varepsilon < \varphi(x) \leqslant (n+1)\varepsilon\}.$$

The expression on the right is Lebesgue's original way of defining his integral.

Conversely, given a nonnegative function φ, suppose all the sets (5) are summable and

$$\sum_{n=1}^{\infty} n\varepsilon\mu(E_{\varepsilon n}) \leqslant C$$

for all $\varepsilon > 0$. Then $\varphi(x)$ is summable. In fact, by Levi's theorem, the function $\varphi_\varepsilon(x)$ defined by (6) is summable for all ε, with integral (7). Since $\varphi_\varepsilon(x) \to \varphi(x)$ as $\varepsilon \to 0$, as already noted, and since $I\varphi_\varepsilon \leqslant C$ for all ε, it follows from Fatou's lemma (see Sec. 2.8.2) that $\varphi(x)$ is summable, with integral (7).

To treat the case where φ is of variable sign, we use the fact that φ is summable if and only if $|\varphi|$ is summable. Therefore φ is summable if and only if

$$\sum_{n=1}^{\infty} n\varepsilon[\mu\{x: n\varepsilon < \varphi(x) \leqslant (n+1)\varepsilon\} + \mu\{x: -(n+1)\varepsilon \leqslant \varphi(x) < -n\varepsilon\} \leqslant C \tag{8}$$

for all $\varepsilon > 0$ (provided the appropriate sets are all summable). Then the integral of $|\varphi|$ is given by the limit as $\varepsilon \to 0$ of the left-hand side of (8), while the integral of φ itself is given by the limit as $\varepsilon \to 0$ of the expression

$$\sum_{n=1}^{\infty} n\varepsilon[\mu\{x: n\varepsilon < \varphi(x) \leqslant (n+1)\varepsilon\} - \mu\{x: -(n+1)\varepsilon \leqslant \varphi(x) < -n\varepsilon\}].$$

---

[3] For simplicity, given a set$\{\cdots\}$, we write $\mu\{\cdots\}$ instead of $\mu(\{\cdots\})$.

6.7. Integration over a Measurable Subset

Until now, the region of integration has been the whole set X. However, it is an easy matter to define integration over an arbitrary measurable subset $E \subset X$. First we note that the product of two measurable functions f and g is again a measurable function. In fact, confining ourselves to nonnegative functions (which obviously entails no loss of generality), we need only note that

$$\{x : f(x)g(x) > c\} = \bigcup_r (\{x : f(x) > r\} \cap \{x : g(x) > c/r\})$$

for any $c > 0$, where r is an arbitrary positive rational number. In particular, the product of any measurable function $f(x)$ with the characteristic function $\chi_E(x)$ of a measurable set E [$f(x)$ replaced by zero outside E] is again a measurable function. Similarly, a summable function replaced by zero outside a measurable set is again summable.

With this in mind, let $\varphi(x)$ be an arbitrary function defined on X. Then we say that $\varphi(x)$ is *summable (measurable) on E* if the product $\chi_E\varphi$ is summable (measurable) on X, and we set

$$\int_E \varphi\, dx = I(\chi_E\varphi),$$

by definition. Obviously, the integral over E has all the ordinary properties of the integral. Moreover, the following special properties are worthy of explicit consideration:

a) *If $|\varphi(x)| \leqslant M$ on E, then*

$$\int_E |\varphi|\, dx \leqslant M\mu(E).$$

In fact, $\chi_E|\varphi| \leqslant M\chi_E$ on X, and hence

$$\int_E |\varphi|\, dx = I(\chi_E\,|\varphi|) \leqslant MI\chi_E = M\mu(E).$$

b) *If φ is summable (measurable) on $E = E_1 \cup E_2 \cup \cdots$, where $E_1, E_2, \ldots$ are disjoint measurable sets, then φ is summable (measurable) on every E_n. Moreover, if φ is summable on E, then*

$$\int_E \varphi\, dx = \int_{E_1} \varphi\, dx + \int_{E_2} \varphi\, dx + \cdots. \tag{9}$$

To see this, we note that if $\chi_E\varphi$ is measurable (summable) on X, then so is $\chi_{E_n}\chi_E\varphi = \chi_{E_n}\varphi$. Moreover, $\chi_{E_1} + \chi_{E_2} + \cdots = \chi_E$, and hence

$$\chi_{E_1}\varphi + \chi_{E_2}\varphi + \cdots = \chi_E\,\varphi.$$

Therefore, if φ is summable on E, the partial sums of the series on the left are bounded by the summable function $\chi_E|\varphi|$. This allows us to integrate term by term, and leads at once to (9).

c) *Given a sequence of measurable sets $E_1, E_2, \ldots$, which are not necessarily disjoint, suppose φ is measurable on every E_n. Then φ is also measurable on $E = E_1 \cup E_2 \cup \cdots$.* In fact,

$$\chi_E\varphi = \chi_{E_1\cup E_2\cup\cdots}\varphi = \chi_{E_1}\varphi + \chi_{E_2-E_1E_2}\varphi + \chi_{E_3-E_1E_3-E_2E_3}\varphi + \cdots,$$

where, by hypothesis and Property b above, every function on the right is measurable. Therefore, by Theorem 2, $\chi_E\varphi$ is measurable, as asserted. Let the E_n be disjoint, let φ be summable on every E_n, and suppose the series in the right-hand side of (9) converges. We would like to conclude that φ is summable and that (9) holds (the converse of Property b). This conclusion is in general false (see Prob. 11, p. 133), but does hold if we impose the extra condition that φ be *nonnegative* on every E_n. Then $\chi_E\varphi$ is the limit of the nondecreasing sequence

$$\chi_{E_1}\varphi + \cdots + \chi_{E_n}\varphi \qquad (n = 1, 2, \ldots),$$

with bounded integrals. Therefore, according to Levi's theorem, $\chi_E\varphi$ is summable and the relation (9) holds. This last fact is sometimes stated in the following equivalent form: *If the function φ is nonnegative and summable on every set $E_1, E_2, \ldots$, where $E_1 \subset E_2 \subset \cdots$ and if*

$$\int_{E_n} \varphi(x)\,dx \leqslant C$$

for all n, then φ is summable on $E = E_1 \cup E_2 \cup \cdots$, and

$$\int_E \varphi(x)\,dx = \lim_{n\to\infty} \int_{E_n} \varphi(x)\,dx.$$

d) *Absolute continuity of the integral on a set.* The integral of a summable function φ on a summable set E approaches zero as $\mu(E) \to 0$, regardless of the character of E. More exactly, given any $\varepsilon > 0$, there exists a $\delta > 0$ such that $\mu(E) < \delta$ implies

$$\left| \int_E \varphi(x)\,dx \right| < \varepsilon.$$

To see this, let $h(x) \geqslant 0$ be an elementary function such that

$$I(|\,|\varphi| - h|) < \frac{\varepsilon}{2}.$$

The function $h(x)$ is bounded, i.e., $0 \leqslant h(x) \leqslant M$ for some M. If the summable set E has measure less than $\delta = \varepsilon/2M$, then

$$\left|\int_E \varphi(x)\,dx\right| \leqslant \int_E |\varphi(x)|\,dx \leqslant \int_E \big|\,|\varphi(x)| - h(x)\big|\,dx$$

$$+\int_E h(x)\,dx < \frac{\varepsilon}{2} + M\delta < \varepsilon,$$

as asserted.

6.8. Measure on a Product Space

In Sec. 2.10 we constructed the Lebesgue integral on the Cartesian product $X \times Y$ of two sets X and Y, starting from a family $H(W)$ of elementary functions $h(x, y)$ satisfying the hypotheses of Fubini's theorem. In Sec. 3.3 the existence of such a family was verified for the special case where X and Y are finite-dimensional blocks. We are now in a position to construct a suitable family $H(W)$ for the case of arbitrary sets X and Y, equipped with Lebesgue integrals I_X and I_Y. In fact, for $H(W)$ we make the "natural choice," i.e., the family of all functions of the form

$$h(x, y) = \sum_{j=1}^{m} \alpha_j \chi_{E_j}(x)\chi_{F_j}(y),$$

where m is arbitrary, $\chi_{E_j}(x)$ is the characteristic function of the set $E_j \subset X$, $\chi_{F_j}(y)$ is the characteristic function of the set $F_j \subset Y$, and all the E_j and F_j are summable. The family $H(W)$ is obviously closed under the formation of linear combinations. Moreover, without loss of generality, we can assume that the sets $E_j \subset X$ are disjoint, and similarly for the sets $F_j \subset Y$. Then, if $h(x, y)$ belongs to $H(W)$, so does its absolute value

$$|h(x, y)| = \sum_{j=1}^{m} |\alpha_j|\, \chi_{E_j}(x)\chi_{F_j}(y).$$

Therefore $H(W)$ satisfies Axioms a and b, p. 23.

Next we define an integral Ih on $H(W)$. Given any $h(x, y) \in H(W)$, let

$$Ih = \sum_{j=1}^{m} \alpha_j \mu_X(E_j)\mu_Y(F_j),$$

where μ_X and μ_Y denote the measures in the spaces X and Y, respectively.[4] The space $H(W)$, equipped with this integral, clearly satisfies all the hypotheses

[4] Thus, for example, $\mu_X(E_j) = I_X\chi_{E_j}(x)$, and similarly for μ_Y.

of Fubini's theorem, i.e., every function $h(x, y) \in H(W)$ is summable in x for all y, the integral

$$I_X h = \sum_{j=1}^{m} \alpha_j \mu_X(E_j) \chi_{F_j}(y)$$

is summable in y, and

$$Ih = \sum_{j=1}^{m} \alpha_j \mu_X(E_j) \mu_Y(F_j) = I_Y\{I_X h(x, y)\}.$$

Moreover, the integral Ih satisfies Axioms 1–3, p. 24:

1) $I(\alpha h + \beta k) = \alpha Ih + \beta Ik$;
2) $Ih \geqslant 0$ if $h(x) \geqslant 0$;
3) $Ih_n \to 0$ if $h_n \searrow 0$.

Axioms 1 and 2 are obvious, while to verify Axiom 3, we need only note that if $h_n(x, y) \searrow 0$ for every x and y, then $I_X h_n(x, y) \searrow 0$ by Levi's theorem, and hence $I_Y\{I_X h_n(x, y)\} = Ih_n \to 0$ by the same theorem. Therefore we can use the Daniell scheme to construct a space $L(W)$ of I-summable functions. Since the space $H(W)$ satisfies all the hypotheses of Fubini's theorem, so does $L(W)$. In particular,

$$I\varphi = I_Y\{I_X \varphi(x, y)\}$$

for every $\varphi(x, y) \in L(W)$. Moreover, we can also write

$$I\varphi = I_X\{I_Y \varphi(x, y)\},$$

because of the symmetry between the roles of x and y in the definition of the elementary integral.

*6.9. The Space L_p

If $f(x)$ is measurable, then so is $|f(x)|^p$ for any $p > 0$, since

$$\{x : |f(x)|^p > C\} = \{x : |f(x)| > C^{1/p}\}$$

for any $C \geqslant 0$. Consider the space $L_p = L_p(X)$ consisting of all measurable functions $f(x)$, defined on a given set X, for which

$$I(|f|^p) = \int_X |f(x)|^p \, dx < \infty.$$

For $p > 0$, L_p is a linear space. In fact, if f belongs to L_p, then obviously so does αf for any real α. Moreover, if $f \in L_p$, $g \in L_p$, then $f + g \in L_p$, since measurability of f and g implies that of $f + g$, and

$$\begin{aligned} |f + g|^p &\leqslant (|f| + |g|)^p \leqslant [2 \sup (|f|, |g|)]^p \\ &= 2^p[\sup (|f|^p, |g|^p)] \leqslant 2^p(|f|^p + |g|^p). \end{aligned}$$

We intend to show that introduction of the (nonnegative) norm

$$\| f \|_p = [I(|f|^p)]^{1/p} \equiv I^{1/p}(|f|^p) \qquad (p \geqslant 1) \tag{10}$$

makes L_p into a *complete* normed linear space. For $p = 1$, this fact has already been proved in Sec. 2.9 (where it was called the Riesz-Fischer theorem). Therefore we can now confine ourselves to the case $p > 1$. The norm (10) obviously satisfies Properties a and b on p. 38:

a) $\| f \|_p > 0$ if $f \neq 0$ (almost everywhere), and $\|0\|_p = 0$.
b) $\| \alpha f \|_p = |\alpha| \, \| f \|_p$ for every $f \in R$ and every real number α.

It is a bit more difficult to prove the triangle inequality:

LEMMA 1. *If* $\eta = \omega(\xi)$ *is an increasing continuous function such that* $\omega(0) = 0$, *and if* $\xi = \lambda(\eta)$ *is the corresponding inverse function (itself automatically continuous and increasing), then*

$$xy \leqslant \int_0^x \omega(\xi)\, d\xi + \int_0^y \lambda(\eta)\, d\eta \tag{11}$$

for arbitrary $x > 0, y > 0$. *In particular*

$$xy \leqslant \frac{x^p}{p} + \frac{x^q}{q} \qquad (p > 1, q > 1), \tag{12}$$

where

$$\frac{1}{p} + \frac{1}{q} = 1.$$

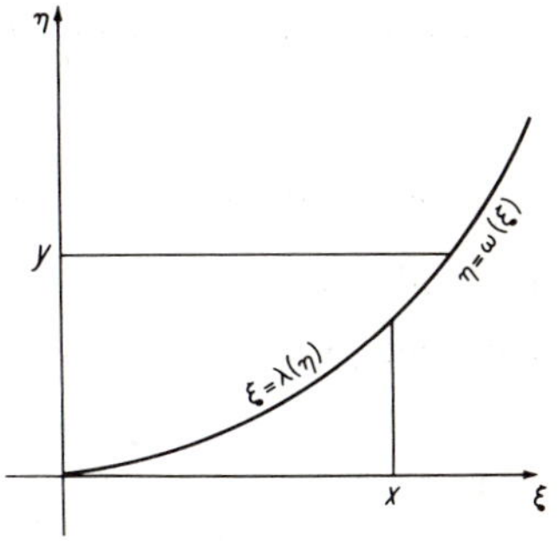

FIGURE 2

Proof. The inequality (11) is geometrically obvious from Figure 2. To deduce (12), substitute $\omega(\xi) = \xi^{p-1}$ $(p > 1)$ and $\lambda(\eta) = \eta^{1/(p-1)}$ into (11), and let

$$q = \frac{1}{p-1} + 1 = \frac{p}{p-1} = \frac{1}{1 - \dfrac{1}{p}}. \tag{13}$$

LEMMA 2 (*Hölder's inequality*). *If* $f \in L_p$, $g \in L_q$, *where*

$$\frac{1}{p} + \frac{1}{q} = 1 \qquad (p > 1, q > 1),$$

then

$$I(|fg|) \leqslant \| f \|_p \, \|g\|_q.$$

Proof. Applying the inequality (12) to the functions $|f(x)|$ and $|g(x)|$, we have

$$|f(x)|\,|g(x)| \leqslant \frac{|f(x)|^p}{p} + \frac{|g(x)|^q}{q}. \tag{14}$$

If

$$\| f \|_p = I^{1/p}(|f|^p) = 1, \qquad \|g\|_q = I^{1/q}(|g|^q) = 1, \tag{15}$$

then, integrating (14), we obtain

$$I(|fg|) \leqslant \frac{1}{p} + \frac{1}{q} = 1. \tag{16}$$

More generally, if (15) is not satisfied, consider the functions

$$f_0 = \frac{f}{\|f\|_p}, \qquad g_0 = \frac{g}{\|g\|_q}$$

instead. Then $\|f_0\|_p = \|g_0\|_q = 1$, and (16) implies

$$I(|f_0 g_0|) = \frac{I(|fg|)}{\|f\|_p \|g\|_q} \leqslant 1,$$

which is equivalent to (14).

THEOREM 6. *The norm* (10) *satisfies the triangle inequality*

$$\|f + g\|_p \leqslant \|f\|_p + \|g\|_p \qquad (f, g \in L_p).$$

Proof. Given that $f, g \in L_p$, we have

$$|f+g|^p \leqslant (|f| + |g|)^p = |f|\,(|f| + |g|)^{p-1} + |g|\,(|f| + |g|)^{p-1}, \tag{17}$$

and hence, since (13) implies $p - 1 = p/q$,

$$(|f| + |g|)^{p-1} = (|f| + |g|)^{p/q} \in L_q,$$

where

$$\|(|f| + |g|)^{p-1}\|_q = I^{1/q}[(|f| + |g|)^p]. \tag{18}$$

Integrating (17), and using Hölder's inequality and (18), we obtain

$$\begin{aligned} I[(|f| + |g|)^p] &\leqslant \|f\|_p \, \|(|f| + |g|)^{p-1}\|_q + \|g\|_p \, \|(|f| + |g|)^{p-1}\|_q \\ &= (\|f\|_p + \|g\|_q) I^{1/q}[(|f| + |g|)^p], \end{aligned}$$

or

$$I^{1/p}[(|f| + |g|)^p] \leqslant \|f\|_p + \|g\|_q,$$

after dividing by $I^{1/q}[(|f| + |g|)^p]$ and recalling that

$$1 - \frac{1}{q} = \frac{1}{p}.$$

But then

$$\|f+g\|_p = I^{1/p}(|f+g|^p) \leqslant I^{1/p}[(|f| + |g|)^p] \leqslant \|f\|_p + \|g\|_p,$$

as required.

THEOREM 7. *The space L_p is complete.*

Proof. The proof differs only slightly from that of the Riesz-Fischer theorem (corresponding to the case $p = 1$). Given a Cauchy sequence $\varphi_n \in L_p$, it is enough to show that φ_n contains a subsequence φ_{n_k} with a limit $\varphi \in L_p$, since then φ will also be the limit of the whole sequence φ_n. This follows from the inequality

$$\|\varphi - \varphi_n\|_p \leqslant \|\varphi - \varphi_{n_k}\|_p + \|\varphi_{n_k} - \varphi_n\|_p$$

and the fact that the second term on the right goes to zero as $n \to \infty$ and $n_k \to \infty$. Clearly, we can always find an increasing sequence of indices n_k such that

$$\|\varphi_n - \varphi_{n_k}\|_p < \frac{1}{2^k} \qquad (k = 1, 2, \ldots)$$

for $n > n_k$. In particular,

$$\|\varphi_{n_{k+1}} - \varphi_{n_k}\|_p < \frac{1}{2^k},$$

which implies that the series

$$\sum_{k=1}^{\infty} |\varphi_{n_{k+1}} - \varphi_{n_k}| \tag{19}$$

converges almost everywhere. In fact,

$$I^{1/p}\left(\sum_{k=1}^{N} |\varphi_{n_{k+1}} - \varphi_{n_k}|\right)^p = \left\|\sum_{k=1}^{N} |\varphi_{n_{k+1}} - \varphi_{n_k}|\right\|_p$$
$$\leqslant \sum_{k=1}^{N} \|\varphi_{n_{k+1}} - \varphi_{n_k}\|_p < \sum_{k=1}^{N} \frac{1}{2^k} < 1,$$

and the assertion then follows by taking the limit as $N \to \infty$ and invoking Levi's theorem. Since (19) converges almost everywhere, the same is true of the series

$$\sum_{k=1}^{\infty} (\varphi_{n_{k+1}} - \varphi_{n_k}),$$

with partial sums

$$\sum_{k=1}^{N} (\varphi_{n_{k+1}} - \varphi_{n_k}) = \varphi_{n_{N+1}} - \varphi_{n_1}.$$

This means that the sequence φ_{n_k} has a limit (almost everywhere) as $k \to \infty$. Let φ denote this limit. Then, for fixed k, the function $\varphi_{n_p} - \varphi_{n_k}$ approaches $\varphi - \varphi_{n_k}$ almost everywhere as $p \to \infty$. Since

$$I^{1/p}(|\varphi_{n_p} - \varphi_{n_k}|^p) = \|\varphi_{n_p} - \varphi_{n_k}\|_p < \frac{1}{2^k} \qquad (p > k),$$

it follows from the result of Sec. 2.8.2 that $\varphi - \varphi_{n_k}$ belongs to L_p, and hence so does φ itself. Moreover, by the same result,

$$\|\varphi - \varphi_{n_k}\|_p = I^{1/p}(|\varphi - \varphi_{n_k}|^p) \leqslant \frac{1}{2^k}.$$

Therefore φ_{n_k} converges to φ in the norm of the space L_p, and the proof is complete.

It is not hard to see that every elementary function $h(x) \in H$ belongs to the space L_p. In fact, suppose $|h(x)| \leqslant M$ (an elementary function must be bounded), and consider the set $E = \{x: |h(x)| > 1\}$. Then E is summable and

$$|h(x)|^p \leqslant \begin{cases} M^p & \text{if } x \in E, \\ |h(x)| & \text{if } x \notin E. \end{cases}$$

In other words,

$$|h(x)|^p \leqslant M^p \chi_E(x) + |h(x)| \chi_{X-E}(x) \leqslant M^p \chi_E(x) + |h(x)|.$$

But both terms on the right are summable, and hence so is $|h(x)|^p$, i.e., $H \subset L_p$, as asserted. Actually, much more can be said:

THEOREM 8. *The family of elementary functions H is dense in L_p.*

Proof. We want to show that any function $f \in L_p$ can be approximated arbitrarily closely in the L_p-norm by an elementary function. Since if f belongs to L_p, so do f^+ and f^-, there is no loss of generality in assuming that f is nonnegative. Consider the increasing sequence of measurable sets

$$E_n = \left\{x: \frac{1}{n} < f(x) \leqslant n\right\} \qquad (n = 1, 2, \ldots),$$

and let

$$f_n(x) = \begin{cases} f(x) & \text{if } x \in E_n, \\ 0 & \text{if } x \notin E_n. \end{cases}$$

Obviously $f_n \nearrow f$ and $(f - f_n)^p \searrow 0$, and hence, by Levi's theorem,

$$\|f - f_n\|_p = I^{1/p}[(f - f_n)^p] \to 0.$$

Therefore, given any $\varepsilon > 0$, we can choose n such that

$$\|f - f_n\| < \frac{\varepsilon}{2}. \tag{20}$$

The set E_n is summable, since $\chi_{E_n} = \chi_{E_n}^p \leqslant n^p f^p$. But then the function f_n is summable, since, by Hölder's inequality,

$$If_n = I(\chi_{E_n} f) \leqslant I^{1/q}(\chi_{E_n}^q)\, I^{1/p}(f^p).$$

Therefore, since H is dense in the space L_1, as shown in Sec. 2.9, there exists a sequence of elementary functions h_k converging to f_n in the L_1-norm (as $k \to \infty$). It can be assumed that the h_k are nonnegative, since the convergence of h_k to f_n is unaffected by replacing h_k by h_k^+. Moreover, it can be assumed that the h_k are bounded by the number n, since the convergence of h_k to f_n is also unaffected by replacing h_k by

$$\min(h_k, n) = n \min\left(\frac{1}{n} h_k, 1\right),$$

where every function on the right is elementary by Stone's Axiom c on p. 119. But then $h_k \to f_n$ in the L_p-norm, as well as in the L_1-norm. In fact,

$$\|f_n - h_k\|_p = I^{1/p}(|f_n - h_k|^p) = I^{1/p}(|f_n - h_k|^{p-1}\,|f_n - h_k|)$$
$$\leqslant n^{(p-1)/p} I^{1/p}(|f_n - h_k|) \to 0 \qquad \text{as} \quad k \to \infty,$$

and hence we can choose k such that

$$\|f_n - h_k\|_p < \frac{\varepsilon}{2}. \tag{21}$$

Combining (20) and (21), we have

$$\|f - h_k\|_p \leqslant \|f_n - h_k\|_p + \|f - f_n\|_p < \varepsilon,$$

and the proof is complete, since ε is arbitrary.

PROBLEMS

1. Prove that a continuous function of one or several measurable functions is measurable.

Hint. A continuous function is a limit of polynomials.

Comment. On the other hand, a measurable function of continuous functions need not be measurable (see Prob. 7, p. 204).

2. Find a sequence of measurable sets $E_1 \supset E_2 \supset \cdots$ for which

$$\mu\left(\bigcap_{n=1}^{\infty} E_n\right) \neq \lim_{n\to\infty} \mu(E_n).$$

Hint. The use of finite $\mu(E_n)$ is precluded by Theorem 4, p. 119. Consider the sets $E_n = \{x: n < x < \infty\}$, $n = 1, 2 \ldots$ with empty intersection and infinite measure.

3. Prove that the symbol $>$ can be replaced by $\geqslant$ in Theorem 5, p. 120.

4 (*Egorov's theorem*). Let $f_1, f_2, \ldots$ be a sequence of measurable functions defined on a summable set X, and suppose f_n converges (almost everywhere) to a function f. Prove that given any $\varepsilon > 0$, there exists a set $E \subset X$ with $\mu(E) > \mu(X) - \varepsilon$ on which f_n converges to f uniformly.

Hint. It can be assumed that $f = 0$ and $f_n \searrow 0$. Consider the sets

$$E_n^{(m)} = \left\{x : 0 < f_n < \frac{1}{m}\right\}.$$

Given any $\varepsilon > 0$, there exists an $n = n(m)$ such that

$$\mu(A_{n(m)}^{(n)}) > \mu(X) - \frac{\varepsilon}{2^m}.$$

Now let

$$E = \bigcap_{m=1}^{\infty} E_{n(m)}^{(m)}.$$

5. Let $f_1, f_2, \ldots$ be a sequence of measurable functions defined on a set X, and let E be the set on which f_n converges. Show that E is measurable.

Hint. $$E = \bigcap_k \bigcup_n \bigcap_m \left\{x : |f_n(x) - f_m(x)| < \frac{1}{k}\right\}.$$

6. Let $f_1, f_2, \ldots$ be a sequence of measurable functions defined on a summable set X, and suppose f_n converges (almost everywhere) to a function f. Show that

$$\lim_{n\to\infty} \mu\{x : |f(x) - f_n(x)| \geqslant c\} = 0 \qquad \text{for arbitrary } c > 0. \tag{22}$$

Hint. The set

$$\bigcap_{m=1}^{\infty} \bigcup_{n=m}^{\infty} \{x : |f(x) - f_n(x)| \geqslant c\}$$

is empty.

Comment. We say that a sequence of measurable functions f_n *converges in measure* to f if it satisfies (22).

7. Show that a sequence of measurable functions converging in measure to a function f always contains a subsequence converging almost everywhere to f, although the sequence itself may not converge almost everywhere to f.

Hint. Given any integers k and m, there exists an $n = n(k, m)$ such that

$$\mu\left\{x : |f_n(x) - f(x)| > \frac{1}{k}\right\} < \frac{1}{2^k m}.$$

As $k \to \infty$, the sequence f_n converges to f on a set of measure greater than

$$\mu(X) - \frac{1}{m}.$$

Now consider the sequence $f_{n(m,m)}$.

8. Prove that if every subsequence of a given sequence of measurable functions contains a subsequence converging almost everywhere to a given function f, then f converges in measure to f.

Hint. Assume the opposite and use Prob. 7.

9. Prove that together $f_n(x) \geqslant 0$ and $If_n \to 0$ imply $f_n \to 0$ in measure, but not $f_n \to 0$ almost everywhere. Show that the condition $f_n(x) \geqslant 0$ cannot be dropped.

Hint. Use Prob. 8.

10. Introduce the metric

$$\rho(f, g) = I\left(\frac{|f-g|}{1+|f-g|}\right) \tag{23}$$

in the space $\mathscr{M}$ of all measurable functions defined on a summable set X. Verify that ρ has all the properties of a metric. Show that convergence in the metric (23) is equivalent to convergence in measure. Show that $\mathscr{M}$ is complete.

11. Given the sets

$$E = (0, 1], \qquad E_n = \left(\frac{1}{2^n}, \frac{1}{2^{n-1}}\right] \qquad (n = 1, 2, \ldots),$$

construct a function $\varphi(x)$ summable on every E_n but not on E, despite the convergence of the series

$$\int_{E_1} \varphi(x)\,dx + \int_{E_2} \varphi(x)\,dx + \cdots.$$

Hint. Suppose that

$$\int_{E_n} \varphi^+(x)\,dx = \int_{E_n} \varphi^-(x)\,dx = 1$$

on every E_n.

7

CONSTRUCTIVE MEASURE THEORY

In this chapter, we describe the approximation of measurable sets by sets of a simpler kind, which in the n-dimensional case are just blocks and their finite and countable combinations. We shall then be able to give a constructive definition of a measurable set and its measure.

7.1. Semirings of Subsets

A family $\mathfrak{A}$ of subsets $A \subset X$ is called a *semiring* if it has the following two properties:

1) If $A \in \mathfrak{A}$, $B \in \mathfrak{A}$, then $AB \in \mathfrak{A}$.
2) If $A_1 \in \mathfrak{A}$, $A \in \mathfrak{A}$ and $A_1 \subset A$, then there exist sets $A_2, \ldots, A_m \in \mathfrak{A}$ such that

$$A = A_1 \cup A_2 \cup \cdots \cup A_m,$$

where the sets $A_1, A_2, \ldots, A_m$ are (pairwise) disjoint.

Example. The set of all subblocks of an n-dimensional block is a semiring. In fact, Properties 1 and 2 have already been stated for blocks on p. 62.

Next we prove two further properties of semirings:

3) Let $A_1, \ldots, A_k$ be k disjoint sets in $\mathfrak{A}$, all contained in a given set $A \in \mathfrak{A}$. Then there exist sets $B_{k+1}, \ldots, B_m \in \mathfrak{A}$ such that

$$A = A_1 \cup \cdots \cup A_k \cup B_{k+1} \cup \cdots \cup B_m, \tag{1}$$

where the sets $A_1, \ldots, A_k, B_{k+1}, \ldots, B_m$ are themselves disjoint. For $k = 1$ this assertion is just Property 2 above. Suppose the decomposition (1) holds for some integer k. Then, as we now show, it also holds for $k + 1$, thereby leading to a proof by induction. In fact, if $A_{k+1} \subset A$ and if A_{k+1} intersects none of the sets $A_1, \ldots, A_k$, then

$$A_{k+1} = A_{k+1}B_{k+1} \cup \cdots \cup A_{k+1}B_m. \tag{2}$$

But, by the definition of a semiring, we have

$$\begin{aligned} B_{k+1} &= A_{k+1}B_{k+1} \cup B_{k+1}^{(1)} \cup \cdots \cup B_{k+1}^{(p_{k+1})}, \\ &\cdots\cdots\cdots\cdots\cdots\cdots \\ B_m &= A_{k+1}B_m \cup B_m^{(1)} \cup \cdots \cup B_m^{(p_m)}, \end{aligned} \tag{3}$$

where the sets $B_{k+1}^{(1)}, \ldots, B_{k+1}^{(p_{k+1})}$ are disjoint, and similarly for $B_m^{(1)}, \ldots, B_m^{(p_m)}$. Substituting (3) into (1) and using (2), we obtain the desired result.

4) The union of an arbitrary finite collection $A_1, \ldots, A_m$ of sets in $\mathfrak{A}$ can be represented in the form

$$A_1 \cup \cdots \cup A_m = A_1^{(1)} \cup \cdots \cup A_1^{(k_1)} \cup \cdots \cup A_m^{(1)} \cup \cdots A_m^{(k_m)}, \tag{4}$$

where the sets on the right are all disjoint and belong to $\mathfrak{A}$, and

$$A_j^{(1)}, \ldots, A_j^{(k_j)} \subset A_j \qquad (j = 1, \ldots, m).$$

For $m = 1$ the assertion is obvious. Suppose the decomposition (4) holds for some integer m. Then, as we now show, it also holds for $m + 1$. In fact, by Property 3,

$$A_{m+1} = A_{m+1}A_1^{(1)} \cup \cdots \cup A_{m+1}A_m^{(k_m)} \cup A_{m+1}^{(1)} \cup \cdots \cup A_{m+1}^{(k_{m+1})}, \tag{5}$$

where the sets on the right are disjoint and belong to $\mathfrak{A}$. But then, combining (4) and (5), we obtain

$$A_1 \cup \cdots \cup A_m \cup A_{m+1}$$
$$= A_1^{(1)} \cup \cdots \cup A_m^{(k_m)} \cup A_{m+1}^{(1)} \cup \cdots \cup A_{m+1}^{(k_{m+1})},$$

where the sets on the right are disjoint and belong to $\mathfrak{A}$, and moreover

$$A_{m+1}^{(1)}, \ldots, A_{m+1}^{(k_{m+1})} \subset A_{m+1},$$

as required. In other words, adding a term A_{m+1} to the union $A_1 \cup \cdots \cup A_m$ leads to the appearance of new terms $A_{m+1}^{(1)}, \ldots, A_{m+1}^{(k_{m+1})}$ without changing the terms originally in the decomposition. Therefore the decomposition also holds for a countable collection of sets $A_1, \ldots, A_m, \ldots$ This fact will be needed later.

7.2. The Subspace Generated by a Semiring of Summable Subsets

Let $\mathfrak{A}$ be a semiring of *summable* subsets of a space X (equipped with a Lebesgue integral I), and let $H_{\mathfrak{A}}$ be the set of all finite linear combinations of characteristic functions of the sets of $\mathfrak{A}$. Then every function

$$h(x) = \sum_{k=1}^{m} \alpha_k \chi_{E_k}(x) \tag{6}$$

in $H_{\mathfrak{A}}$ has a well-defined integral

$$Ih = \sum_{k=1}^{m} \alpha_k \mu(E_k). \tag{7}$$

This immediately suggests the following question: Can we use the Daniell scheme to construct an integral (as in Chap. 2), starting from the set $H_{\mathfrak{A}}$ and the integral (7), and if so, what does the construction give?

According to Property 4, the sets E_k figuring in (6) can always be regarded as disjoint, so that

$$|h(x)| = \sum_{k=1}^{m} |\alpha_k| \, \chi_{E_k}(x)$$

again belongs to $H_{\mathfrak{A}}$. Moreover, the set $H_{\mathfrak{A}}$ is obviously closed under the formation of linear combinations. Therefore $H_{\mathfrak{A}} = H$ satisfies Axioms a and b for a family of elementary functions (see p. 23). Furthermore, the integral (7) satisfies Axioms 1–3 for an elementary integral. In fact, Axioms 1 and 2 are obvious, while, to verify Axiom 3, we merely note that $h_n \searrow 0$ implies $Ih_n \to 0$, by Levi's theorem. Therefore all the prerequisites for constructing an integral are satisfied. Let us now see what we obtain from this construction.

Suppose the sequence $h_n \in H_{\mathfrak{A}}$ is nondecreasing and has bounded integrals Ih_n. Then the function $f = \lim h_n$ is summable, by Levi's theorem. Therefore the class $L_{\mathfrak{A}}^+$, obtained from $H_{\mathfrak{A}}$ by the construction of Sec. 2.3, is contained in $L(X)$. Completing the construction of the integral by taking differences of functions in $L_{\mathfrak{A}}^+$ (as in Sec. 2.5), we arrive at a class $L_{\mathfrak{A}}$, which in turn must be contained in $L(X)$. Moreover, according to Sec. 2.9, the class $L_{\mathfrak{A}}$ is complete in the $I(|\varphi|)$ norm, and hence is a closed subspace of $L(X)$. On the other hand, as we know from the same section, the elementary functions $h \in H_{\mathfrak{A}}$ are dense in $L_{\mathfrak{A}}$. It follows that *the class $L_{\mathfrak{A}}$ is the closure of the set $H_{\mathfrak{A}}$ in the $I(|\varphi|)$ norm.*

7.3. Sufficient Semirings

Consider the set H_0 of all finite linear combinations of characteristic functions of *all* summable subsets of a given set X. Then it is easy to see that

H_0 is dense in $L(X)$. In fact, we need only verify that every nonnegative function $\varphi \in L(X)$ belongs to $\bar{H}_0$, the closure of H_0 in the $I(|\varphi|)$ norm. Let

$$\varphi_\varepsilon = \sum_{n=1}^{\infty} n\varepsilon\chi_{\varepsilon n}(x) = \sum_{n=1}^{N} n\varepsilon\chi_{\varepsilon n}(x) + \sum_{n=N+1}^{\infty} n\varepsilon\chi_{\varepsilon n}(x)$$

be the same function as in Sec. 6.6. Then the first term on the right belongs to H_0, while the second converges in norm to 0 as $N \to \infty$, since

$$\left\| \sum_{n=N+1}^{\infty} n\varepsilon\chi_{\varepsilon n}(x) \right\| = \sum_{n=N+1}^{\infty} n\varepsilon\mu(E_n) \to 0 \qquad \text{as} \quad N \to \infty.$$

Therefore φ_ε belongs to $\bar{H}_0$. But then φ also belongs to $\bar{H}_0$, as asserted, since $\varphi_\varepsilon \nearrow \varphi$, which implies

$$I(\varphi - \varphi_\varepsilon) = \|\varphi - \varphi_\varepsilon\| \to 0.$$

In other words, according to the last remark of the preceding section, $L_{H_0} = L(X)$.

We now ask the following question: When does $L_{\mathfrak{A}}$ coincide with $L(X)$, or equivalently, when is the semiring $\mathfrak{A}$ (of summable subsets of X) *sufficient* in the sense that linear combinations of characteristic functions of its sets are dense in $L(X)$? The answer to this question is given by

THEOREM 1. *A semiring $\mathfrak{A}$ of summable sets is sufficient if and only if, given any summable set E and any $\varepsilon > 0$, there exists a set F, which is the union of a finite number of sets of $\mathfrak{A}$, such that*

$$\mu(E - EF) + \mu(F - EF) < \varepsilon. \tag{8}$$

Proof. If (8) holds, then

$$I(|\chi_E - \chi_F|) = \|\chi_E - \chi_F\| < \varepsilon,$$

and hence the characteristic function of any summable set E is a limit (in the L-norm) of linear combinations of characteristic functions of sets of the semiring $\mathfrak{A}$. But then linear combinations of characteristic functions of sets of $\mathfrak{A}$ are dense in L.

Conversely, suppose we know that linear combinations of characteristic functions of sets of $\mathfrak{A}$ are dense in L. Then, if E is any summable set,

$$\chi_E(x) = \lim_{n\to\infty} g_n(x)$$

for some sequence

$$g_n(x) = \sum_{k=1}^{r_n} \alpha_{kn}\chi_{E_{kn}}(x), \qquad E_{kn} \in \mathfrak{A}.$$

It can be assumed that the sets E_{kn} $(k = 1, \ldots, r_n)$ are disjoint for every fixed n. Consider the function

$$\hat{g}_n(x) = \sum_{k=1}^{r_n}{}' \chi_{E_{kn}}(x),$$

where the sum is taken only over sets E_{kn} such that $\alpha_{kn} \geqslant \frac{1}{2}$. Writing

$$E_n = E_{1n} \cup \cdots \cup E_{r_n n},$$

we distinguish four possibilities:

1) $\chi_E(x) = \hat{g}_n(x) = 1$ if $x \in EE_n$;
2) $\chi_E(x) = 1$, $\hat{g}_n(x) = 0$, $g_n(x) \leqslant \frac{1}{2}$,
   $|\chi_E(x) - \hat{g}_n(x)| = 1 \leqslant 2\,|\chi_E(x) - g_n(x)|$ if $x \in E(X - E_n)$;
3) $\chi_E(x) = 0$, $\hat{g}_n(x) = 1$, $g_n(x) \geqslant \frac{1}{2}$,
   $|\chi_E(x) - \hat{g}_n(x)| = 1 \leqslant 2\,|\chi_E(x) - g_n(x)|$ if $x \in (X - E)E_n$;
4) $\chi_E(x) = 0$, $\hat{g}_n(x) = 0$ if $x \in (X - E)(X - E_n)$.

Thus

$$|\chi_E(x) - \hat{g}_n(x)| \leqslant 2\,|\chi_E(x) - g_n(x)|$$

for all $x \in X$, and hence

$$\|\chi_E - \hat{g}_n\| = I(|\chi_E - \hat{g}_n|) \leqslant 2I(|\chi_E - g_n|) \to 0 \qquad (9)$$

as $n \to \infty$.

The function $\hat{g}_n(x)$ is itself the characteristic function of some set B_n, which is a finite union of sets of the semiring $\mathfrak{A}$. We now assert that the condition (8) holds, with $F = B_n$, if n is sufficiently large. To see this, we note that $(\chi_E - \hat{g}_n)^+$ is the characteristic function of the set $E - EB_n$, while $(\chi_E - \hat{g}_n)^-$ is the characteristic function of the set $B_n - EB_n$. But then, according to (9),

$$\begin{aligned}\mu(E - EB_n) + \mu(B_n - EB_n) &= I(\chi_E - \hat{g}_n)^+ + I(\chi_E - \hat{g}_n)^- \\ &= I(|\chi_E - \hat{g}_n|) = \|\chi_E - g_n\| \to 0\end{aligned}$$

as $n \to \infty$, and the proof is complete.

Given a family of sets $\mathfrak{A}$, the family of all sets obtained by forming countable unions of sets of $\mathfrak{A}$ will be denoted by $\mathfrak{A}_\sigma$, while the family of all sets obtained by forming countable intersections of sets of $\mathfrak{A}$ will be denoted by $\mathfrak{A}_\delta$. Then we write $\mathfrak{A}_{\sigma\delta} = (\mathfrak{A}_\sigma)_\delta$, $\mathfrak{A}_{\sigma\delta\sigma} = (\mathfrak{A}_{\sigma\delta})_\sigma$, and so on.

LEMMA 1. *If $\mathfrak{A}$ is a sufficient semiring, then, given any summable set E and any $\varepsilon > 0$, there exists a set $F \in \mathfrak{A}_\sigma$ such that*

$$\mu(E - EF) = 0, \qquad \mu(F - EF) < \varepsilon. \qquad (10)$$

Proof. Given any integers m and n, we use Theorem 1 to find a set F_{mn}, which is a finite union of sets of $\mathfrak{A}$, such that

$$\mu(E - EF_{mn}) < \frac{1}{2^n m}, \qquad \mu(F_{mn} - EF_{mn}) < \frac{1}{2^n m}.$$

Then the set

$$F_m = \bigcup_n F_{mn}$$

belongs to $\mathfrak{A}_\sigma$, and moreover

$$E - EF_m = \bigcap_m (E - EF_{mn}), \qquad F_m - EF_m = \bigcup_n (F_{mn} - EF_{mn}).$$

Therefore

$$\mu(E - EF_m) \leqslant \inf_n \mu(E - EF_{mn}) = 0,$$

$$\mu(F_m - EF_m) \leqslant \sum_{n=1}^{\infty} (F_{mn} - EF_{mn}) < \frac{1}{m},$$

and (10) is satisfied by choosing F to be any F_m with $m > 1/\varepsilon$.

THEOREM 2. *If $\mathfrak{A}$ is a sufficient semiring, then, given any summable set E, there exists a set $F \in \mathfrak{A}_{\sigma\delta}$ such that*

$$\mu(E - EF) = \mu(F - EF) = 0.$$

Proof. According to Lemma 1, given any integer m, there exists a set $F_m \in \mathfrak{A}_\sigma$ such that

$$\mu(E - EF_m) = 0, \qquad \mu(F_m - EF_m) < \frac{1}{m}.$$

Then the set

$$F = \bigcap_m F_m$$

belongs to $\mathfrak{A}_{\sigma\delta}$, and moreover

$$E - EF = \bigcup_m (E - EF_m), \qquad F - EF = \bigcap_m (F_m - EF_m).$$

Therefore

$$\mu(E - EF) \leqslant \sum_m \mu(E - EF_m) = 0,$$

$$\mu(F - EF) \leqslant \inf_m \mu(F_m - EF_m) = 0,$$

and the theorem is proved.

Remark. Thus any summable set E can be approximated to within a set of measure zero by forming countable unions and intersections of sets of a sufficient semiring. This follows at once from Theorem 2, since we can always write

$$E = F \cup (E - EF) - (F - EF).$$

7.4. Completely Sufficient Semirings

A sufficient semiring $\mathfrak{A}$ is said to be *completely sufficient* if, given any set Z of measure zero and any $\varepsilon > 0$, there exists a set $A \in \mathfrak{A}_\sigma$ such that $Z \subset A$ and $\mu(A) < \varepsilon$. A sufficient semiring need not be completely sufficient, as shown by the example where $\mathfrak{A}$ is the family of all measurable subsets $E \subset X$ which do not contain a fixed point x_0 of measure zero. Both Lemma 1 and Theorem 2 can be strengthened somewhat if the semiring $\mathfrak{A}$ is completely sufficient. In fact, it turns out that not only can any summable set E be approximated to within a set of measure zero by a set G of the family $\mathfrak{A}_{\sigma\delta}$, but the approximating set G can be chosen to *cover* E ($G \supset E$).

LEMMA 2. *If $\mathfrak{A}$ is a completely sufficient semiring, then, given any summable set E and any $\varepsilon > 0$, there exists a set $G \in \mathfrak{A}_\sigma$ such that*

$$G \supset E, \qquad \mu(G) < \mu(E) + \varepsilon.$$

Proof. According to Lemma 1, given any $\varepsilon > 0$, there exists a set $F \in \mathfrak{A}_\sigma$ such that

$$\mu(E - EF) = 0, \qquad \mu(F - EF) < \frac{\varepsilon}{2}.$$

Then

$$E = F \cup Z - B,$$

where $Z = E - EF$ is a set of measure zero and $B = F - EF$ is a set of measure less than $\varepsilon/2$. In particular,

$$\mu(F) \leqslant \mu(E) + \mu(B) < \mu(E) + \frac{\varepsilon}{2}.$$

Since Z is a set of measure zero and $\mathfrak{A}$ is completely sufficient, Z can be covered by a set $A \in \mathfrak{A}_\sigma$ of measure less than $\varepsilon/2$. Then the set $G = F \cup A$ obviously satisfies all the requirements of the lemma.

THEOREM 3. *If $\mathfrak{A}$ is a completely sufficient semiring, then, given any summable set E, there exists a set $G \in \mathfrak{A}_{\sigma\delta}$ such that*

$$G \supset E, \qquad \mu(G) = \mu(E).$$

Proof. According to Lemma 2, given any integer m, there exists a set $G_m \in \mathfrak{A}_\sigma$ such that

$$G_m \supset E, \qquad \mu(G_m) < \mu(E) + \frac{1}{m}.$$

Then the set

$$G = \bigcap_m G_m$$

contains E and belongs to $\mathfrak{A}_{\sigma\delta}$, and moreover

$$\mu(E) \leqslant \mu(G) \leqslant \mu(G_m) < \mu(E) + \frac{1}{m}.$$

Letting $m \to \infty$, we find that $\mu(G) = \mu(E)$, as asserted. Alternatively, we can write

$$G = E \cup Z,$$

where Z is a set of measure zero.

There is an analogue of Theorem 3 suitable for *measurable* sets:

THEOREM 4. *If $\mathfrak{A}$ is a completely sufficient semiring, then, given any measurable set E, there exists a set $G \in \mathfrak{A}_{\sigma\delta\sigma}$ such that*

$$G \supset E, \qquad \mu(G) = \mu(E). \tag{11}$$

Proof. Let $X_1 \subset X_2 \subset \cdots$ be an increasing sequence of summable sets whose union equals the space X (see Remark 2, p. 121). Then every EX_n is summable, and obviously

$$E = EX_1 \cup EX_2 \cup \cdots.$$

According to Theorem 3, there exists a sequence of sets $G_m \in \mathfrak{A}_{\sigma\delta}$ such that

$$EX_m = G - Z_m \qquad (m = 1, 2, \ldots),$$

where every Z_m is of measure zero. But then

$$E = \bigcup_m (G_m - Z_m) = G - Z, \tag{12}$$

where

$$G = \bigcup_m G_m,$$

and

$$Z = \bigcup_m Z_m$$

is of measure zero. Since (12) and (11) are equivalent, the theorem is proved.

7.5. Outer Measure and the Measurability Criterion

In this section, sets belonging to an underlying completely sufficient semiring $\mathfrak{A}$ of summable sets $E \subset X$ will be called *simple* sets. Given an arbitrary set $E \subset X$, let

$$\mu^*(E) = \inf_{E \subset A_1 \cup A_2 \cup \cdots} \sum_{m=1}^{\infty} \mu(A_m), \tag{13}$$

where the greatest lower bound is taken with respect to all countable coverings of E by simple sets $A_1, A_2, \ldots$ The number $\mu^*(E)$ is called the *outer* (or *upper*) *measure* of E, and the cases where $\mu^*(E) = \infty$ or $\mu^*(E)$ fails to exist (E may have no covering of the indicated type) are not excluded. It follows from the very definition of a completely sufficient semiring that every set of measure zero also has outer measure zero.

THEOREM 5. *If E is summable, then $\mu^*(E)$ exists and equals $\mu(E)$.*

Proof. There is no loss of generality in assuming that the sets $A_1, A_2, \ldots$ figuring in (13) are disjoint, since otherwise, according to Property 4, p. 135, we can successively replace every A_m by disjoint subsets $A_m^{(1)}, \ldots, A_m^{(k_m)}$, whose union equals A_1 if $m = 1$ and $A_m - A_m A_{m-1}$ if $m > 1$. But the sum of the measures of the disjoint sets $A_m^{(j)}$ ($j \leqslant k_m$, $m = 1, 2, \ldots$) cannot exceed the sum of the measures of the original sets A_m ($m = 1, 2, \ldots$). Assuming that the sets $A_1, A_2, \ldots$ are disjoint, we can write (13) in the form

$$\mu^*(E) = \inf_{E \subset A_1 \cup A_2 \cup \cdots} \mu(A_1 \cup A_2 \cup \cdots),$$

using the measurability of $A_1 \cup A_2 \cup \cdots$ and countable additivity. But obviously $\mu(A_1 \cup A_2 \cup \cdots) \geqslant \mu(E)$, since $A_1 \cup A_2 \cup \cdots \supset E$. It follows that

$$\mu^*(E) \geqslant \mu(E), \tag{14}$$

provided $\mu^*(E)$ exists. On the other hand, by Lemma 2, there exists a sequence of sets $G_m \in \mathfrak{A}_\sigma$ such that

$$G_m \supset E, \qquad \mu(G_m) < \mu(E) + \frac{1}{m} \qquad (m = 1, 2, \ldots).$$

Since G_m is a countable covering of E by simple sets, $\mu^*(E)$ exists and moreover

$$\mu^*(E) \leqslant \mu(G_m) < \mu(E) + \frac{1}{m}.$$

Taking the limit as $m \to \infty$, we obtain

$$\mu^*(E) \leqslant \mu(E), \tag{15}$$

and the theorem now follows by comparing (14) and (15).

It is now natural to ask whether measurable sets can be defined directly in terms of outer measure. In the case where the space X is summable, the answer is given by

THEOREM 6 (*Measurability criterion*). *If X is summable, then the set $E \subset X$ is measurable if and only if*

$$\mu^*(E) + \mu^*(\mathscr{C}E) = \mu(X), \tag{16}$$

where $\mathscr{C}E = X - E$.

Proof. In other words, a necessary and sufficient condition for measurability of E is that the sum of the outer measures of E and its complement $\mathscr{C}E$ (relative to X) be equal to the measure of the whole space X. The necessity of (16) is almost obvious, since if E is measurable, so is $\mathscr{C}E$, and then (16) follows from Theorem 5 and the relation

$$\mu(E) + \mu(\mathscr{C}E) = \mu(X).$$

To prove the sufficiency, suppose E satisfies (16). Then, given any integer m, there exist sets $G_E^{(m)}, G_{\mathscr{C}E}^{(m)} \in \mathfrak{A}_\sigma$ such that $G_E^{(m)} \supset E$, $G_{\mathscr{C}E}^{(m)} \supset \mathscr{C}E$ and

$$\mu(G_E^{(m)}) + \mu(G_{\mathscr{C}E}^{(m)}) < \mu(X) + \frac{1}{m} \qquad (m = 1, 2, \ldots). \tag{17}$$

Let $h_E^{(m)}$ and $h_{\mathscr{C}E}^{(m)}$ be the characteristic functions of the sets $G_E^{(m)}$ and $G_{\mathscr{C}E}^{(m)}$. Then it is easy to see that

$$0 \leqslant 1 - h_{\mathscr{C}E}^{(m)}(x) \leqslant \chi_E(x) \leqslant h_E^{(m)}(x),$$

where χ_E is the characteristic function of E, and hence

$$0 \leqslant I(1 - h_{\mathscr{C}E}^{(m)}) \leqslant Ih_E^{(m)}.$$

On the other hand, (17) implies

$$\begin{aligned} Ih_E^{(m)} - I(1 - h_{\mathscr{C}E}^{(m)}) &= Ih_E^{(m)} + Ih_{\mathscr{C}E}^{(m)} - \mu(X) \\ &= \mu(G_E^{(m)}) + \mu(G_{\mathscr{C}E}^{(m)}) - \mu(X) < \varepsilon. \end{aligned}$$

Clearly, the sequence $h_E^{(m)}$ can be regarded as nonincreasing, and the sequence $1 - h_{\mathscr{C}E}^{(m)}$ as nondecreasing. Therefore

$$h_E^{(m)} - (1 - h_{\mathscr{C}E}^{(m)}) \searrow f,$$

where f is nonnegative and summable with $If = 0$, by Corollary 1 to Levi's theorem. It follows that $f(x) = 0$ almost everywhere, by Corollary 2 to Levi's theorem. But

$$h_E^{(m)}(x) - [1 - h_{\mathscr{C}E}^{(m)}(x)] \geqslant h_E^{(m)}(x) - \chi_E(x) \geqslant 0,$$

and hence

$$\chi_E(x) = \lim_{m\to\infty} h_E^{(m)}(x)$$

almost everywhere, i.e., χ_E is measurable and the proof is complete.

7.6. Measure Theory in n-Space. Examples

We now examine what the above general theory gives when the set X is an n-dimensional block $\mathbf{B}$. Let $\sigma(B)$ be a continuous nonnegative quasi-volume defined on the semiring $\mathfrak{A}$ of all subblocks $B \subset \mathbf{B}$. As in Sec. 5.8, we construct a space $L_\sigma(\mathbf{B})$ of σ-summable functions, equipped with a Lebesgue-Stieltjes integral I_σ, starting from step functions as elementary functions and the quasi-volume $\sigma(B)$ as elementary integral. Then functions which are (σ-almost-everywhere) limits of sequences of step functions are said to be *σ-measurable*, and sets with σ-measurable characteristic functions are themselves said to be σ-measurable. Since $\mathbf{B}$ is finite, every σ-measurable set is automatically σ-summable. Let χ_E be the characteristic function of a σ-measurable set $E \subset \mathbf{B}$. Then the *σ-measure* of E is the quantity $\sigma(E) = I_\sigma\chi_E$, which obviously reduces to the quasi-volume of E if E is a block. According to Chap. 6, the family of σ-measurable sets $E \subset \mathbf{B}$ is closed under the formation of complements and countable unions and intersections. Moreover, σ-measure is countably additive, in the sense of Theorem 3, p. 117.

We would now like to give a constructive characterization of σ-measurable sets, using the theory developed in this chapter. First we note that the family of step functions H is just the family of all finite linear combinations of characteristic functions of sets of the semiring $\mathfrak{A}$. But H is dense in $L_\sigma(\mathbf{B})$, by the Riesz-Fischer theorem. Therefore $\mathfrak{A}$ is a sufficient semiring. It follows from Theorem 1 that given any σ-measurable set $E \subset \mathbf{B}$ and any $\varepsilon > 0$, there is a finite union of blocks $F = B_1 \cup \cdots \cup B_m$ such that

$$\sigma(E - EF) + \sigma(F - EF) < \varepsilon,$$

i.e., to within a set of arbitrarily small σ-measure, every σ-measurable set is a finite union of blocks. Moreover, if Z has σ-measure zero relative to the integral I_σ (cf. footnote 1, p. 89), then, by exactly the same argument as on pp. 14–15, Z can be covered by a countable collection of blocks $B_1, B_2, \ldots$ whose quasi-volumes have a sum less than ε.[1] Therefore the semiring $\mathfrak{A}$ is also *completely sufficient*. This allows us to apply all the results of Secs. 7.4 and 7.5, which in the present context take the following form:

1) *Every σ-measurable set is a countable intersection of countable unions of blocks minus a set of σ-measure zero* (Theorem 3).
2) *The outer measure of a set $E \subset \mathbf{B}$ is the quantity*

$$\sigma^*(E) = \inf_{E\subset B_1\cup B_2\cup\cdots} \sigma(B_1 \cup B_2 \cup \cdots),$$

and $\sigma(E) = \sigma^*(E)$ if E is σ-measurable (Theorem 5).
3) *The set $E \subset \mathbf{B}$ is σ-measurable if and only if*

$$\sigma^*(E) + \sigma^*(\mathbf{B} - E) = \sigma(\mathbf{B})$$

(Theorem 6).

[1] Here, however, we do not require every point of Z to be an interior point of a block, i.e., a covering by a collection of blocks has the usual set-theoretic meaning (contrary to p. 13).

DEFINITION. *A set obtained from blocks by forming no more than a countable number of unions, intersections and complements is called a (classical) Borel set.*

Thus every Borel set is σ-measurable, and every σ-measurable set is the difference between a Borel set and a set of σ-measure zero. (There are sets of σ-measure zero which are not Borel sets, as shown in Prob. 2, p. 148). Every open set G (i.e., every set consisting entirely of interior points) is a Borel set and hence σ-measurable. In fact, every point of G can be covered by a block $B \subset G$ whose boundary sheets have rational coordinates. Similarly, every set F which is closed (relative to $\mathbf{B}$) is a Borel set, being the complement of an open set.

The three examples considered previously in Secs. 4.4 and 5.2 will now be examined from the standpoint of σ-measure:

Example 1. If the quasi-volume $\sigma(B)$ of the block B is its ordinary volume $s(B)$, then the σ-measurable sets are called *Lebesgue measurable* (or just *measurable*), and the quantity $s(E)$ is called the *Lebesgue measure* of E (or just the *measure* of E). All the results of Secs. 7.4 and 7.5, involving outer measure and approximation by countable collections of blocks, are valid for Lebesgue measure.

Example 2. Next we find the structure of the σ-measurable sets generated by the quasi-volume

$$\sigma(B) = \int_B g(x)\, dx,$$

where the function $g(x) \geqslant 0$ is Lebesgue summable over the basic block $\mathbf{B}$. *Every Borel set G is σ-measurable*, and in fact

$$\sigma(G) = I_\sigma \chi_G = I(\chi_G g) = \int_G g(x)\, dx, \tag{18}$$

according to formula (1), p. 91 and Theorem 6, p. 107. *Every set Z of Lebesgue measure zero is σ-measurable*, with $\sigma(Z) = 0$. To see this, let G be a Borel set such that $G \supset Z$, $\mu(G) = 0$. Then, according to (18),

$$\sigma(G) = \int_G g(x)\, dx = 0,$$

and hence $\sigma(Z) \leqslant \sigma(G) = 0$. *Every Lebesgue-measurable set E is σ-measurable*, being the difference between a Borel set and a set of Lebesgue measure zero, and clearly

$$\sigma(E) = \int_E g(x)\, dx.$$

In particular, the set $G_0 = \{x : g(x) = 0\}$ is σ-measurable, and

$$\sigma(G_0) = \int_{G_0} g(x)\, dx = 0.$$

Moreover, *every subset* $Q \subset G_0$ *also has* σ*-measure zero* (although it may not be Lebesgue measurable!). Therefore *the union of a Lebesgue-measurable set and a set* $Q \subset G_0$ *is* σ*-measurable.* In fact, the converse is true, i.e., *every* σ*-measurable set* $E \subset \mathbf{B}$ *is the union of a Lebesgue-measurable set and a set* $Q \subset G_0$. To see this, suppose E is σ-measurable. Then its characteristic function χ_E is σ-summable. Hence, as shown on p. 91, the function $\chi_E g$ is Lebesgue summable, and

$$\sigma(E) = I_\sigma \chi_E = I(\chi_E g) = \int_{\mathbf{B}} \chi_E(x) g(x)\, dx.$$

The set $G_+ = \{x : g(x) > 0\}$ is Lebesgue measurable, and obviously

$$E = EG_0 \cup EG_+. \tag{19}$$

Since the set $EG_+ = \{x : \chi_E(x) g(x) > 0\}$ is also Lebesgue measurable, (19) represents E as the union of a Lebesgue-measurable set and a set on which $g(x)$ vanishes, as required.

On p. 91 it was shown that if φ is σ-summable, then the product φg is Lebesgue summable. We are now in a position to prove the converse (as promised), i.e., *if the product* φg *is Lebesgue summable, then* φ *is* σ*-summable.* First we show that φ is σ-measurable. Given any C, let $E_C(\varphi)$ be the set of points where the inequality $\varphi(x) \leqslant C$ holds. This set coincides with the set A where the inequality $\varphi(x)g(x) \leqslant Cg(x)$ holds, except possibly for a set $A_0 \subset A$ on which $g(x)$ vanishes. The set A is Lebesgue measurable and hence σ-measurable, while A_0 has σ-measure zero. Therefore $E_C(\varphi)$ is σ-measurable, and hence φ is σ-measurable, since C is arbitrary. To prove that φ is σ-summable, we need only show that the integrals $I_\sigma \varphi_m$, where $\varphi_m(x) = \min\{|\varphi(x)|, m\}$, form a bounded sequence. The function φ_m is σ-measurable (since φ is) and bounded. Therefore φ_m is σ-summable, and moreover

$$I_\sigma \varphi_m = I(\varphi_m g) \leqslant I(|\varphi|\, g),$$

as asserted. Thus the class L_σ of σ-summable functions has now been completely characterized: *A function* φ *belongs to* L_σ *if and only if the product* φg *is summable in the ordinary sense.*

Example 3. Consider the σ-measure generated by the quasi-volume

$$\sigma(B) = \sum_{c_m \in B} g_m,$$

where $c_1, \ldots, c_m, \ldots$ is a sequence of points in the basic block $\mathbf{B}$, and $g_1, \ldots, g_m$, is a corresponding sequence of real numbers such that

$$\sum_{m=1}^{\infty} g_m < \infty \qquad (g_m > 0)$$

(cf. Example 3, p. 91). Then *every* set $E \subset \mathbf{B}$ is σ-measurable, with σ-measure given by the formula

$$\sigma(E) = I_\sigma \chi_E = \sum_{c_m \in E} g_m.$$

In fact, E differs from $E_0 \subset E$, the set of points c_m contained in E, only by a set of σ-measure zero. But E_0 is σ-measurable, since only countably many points c_m lie in E_0. Therefore E itself is σ-measurable, and moreover

$$\sigma(E) = \sigma\left(\bigcup_{c_m \in E} \{c_m\}\right) = \sum_{c_m \in E} g_m,$$

as asserted.[2]

7.7. Lebesgue Measure for $n = 1$. Inner Measure

Finally we consider in more detail the simplest case $n = 1$, where $\mathbf{B} = [a, b]$ and μ is Lebesgue measure. As is well known, every open set $G \subset \mathbf{B}$ is the union of a countable number of disjoint open intervals G_j (the *components* of G). Therefore G is measurable, with measure equal to the sum of the lengths of the intervals G_j. But the half-open intervals $(\alpha, \beta] \subset \mathbf{B}$ form a completely sufficient semiring of summable subsets of $\mathbf{B}$, and obviously $\mu(\alpha, \beta] = \beta - \alpha$. Therefore we can define the outer measure of an arbitrary set $E \subset \mathbf{B}$ as the quantity

$$\mu^*(E) = \inf_{E \subset G} \mu(G),$$

where the greatest lower bound is taken with respect to all open sets G containing E. Then, according to Theorem 6, a set $E \subset \mathbf{B}$ is measurable if and only if

$$\mu^*(E) + \mu^*(\mathscr{C}E) = \mu(\mathbf{B}) = b - a. \tag{20}$$

The sets satisfying (20) are precisely those called *measurable* by Lebesgue, and used as the starting point of his theory of measure and integration.

There is another, equivalent definition of measurable sets $E \subset \mathbf{B}$, which is worth mentioning at this point. First we introduce the concept of the *inner* (or *lower*) *measure* of a set E, defined as the quantity

$$\mu_*(E) = \sup_{F \subset E} \mu(F),$$

where the least upper bound is taken with respect to all closed sets F contained in E, and $\mu(F)$ is the measure of F.[3]

---

[2] By $\{c_m\}$ we mean the set whose only element is c_m.

[3] As already noted on p. 145, every closed set F is measurable, being the complement of an open set.

THEOREM 7. *The set* $E \subset \mathbf{B} = [a, b]$ *is measurable if and only if*

$$\mu_*(E) = \mu^*(E). \tag{21}$$

Proof. Obviously, we have

$$\begin{aligned} \mu_*(E) &= \sup_{F \subset E} \mu(F) = \sup_{F \subset E} [b - a - \mu(\mathscr{C}F)] \\ &= b - a - \inf_{F \subset E} \mu(\quad F) = b - a - \inf_{\mathscr{C}E \subset \mathscr{C}F} \mu(F). \end{aligned}$$

But F is closed and contained in E, and hence $\mathscr{C}F$ is open (relative to $\mathbf{B}$) and contains $\mathscr{C}E$. Since $\mathscr{C}F$ is an arbitrary open set, it follows that

$$\mu_*(E) + \mu^*(\mathscr{C}E) = b - a.$$

Therefore (20) and (21) are equivalent, as asserted.

PROBLEMS

1. Show that the family of all Borel sets in the interval [0, 1] has the power of the continuum.

Hint. A Borel set can be specified by using a sequence of closed sets.[4]

2. Show that the family of all Lebesgue-measurable sets in the interval [0, 1], even just sets of measure zero, has a power greater than the power of the continuum.

Hint. Any subset of a set of measure zero is also of measure zero. Use Prob. 2, p. 21.

Comment. Problems 1 and 2 show that there exist measurable sets which are not Borel sets.

3. Given a family $\mathscr{F}$ of measurable sets, no two of which differ by more than a set of measure zero, show that the power of $\mathscr{F}$ is no greater than the power of the continuum.

Hint. Every measurable set is a Borel set to within a set of measure zero.

4. Construct a measurable set E which has positive measure both on every interval and on the complement of every interval.

Hint. Construct the analogue of the Cantor set but with positive measure, and then repeat the construction in every interval adjacent to E (see Probs. 2 and 3, pp. 21–22).

---

[4] For further details, see F. Hausdorff, *Mengenlehre*, third edition, Walter de Gruyter & Co., Berlin (1935), p. 181.

Comment. The function $\chi_E(x)$ is not Riemann integrable, and cannot be made Riemann integrable by modification on any set of measure zero.

5. Prove the following result due to N. N. Luzin: Given any $\varepsilon > 0$ and any measurable function $f(x)$ defined on a finite block $\mathbf{B}$, there exists a closed set $F \subset \mathbf{B}$, with measure $\mu(F) > \mu(\mathbf{B}) - \varepsilon$, on which $f(x)$ is continuous.

Hint. For step functions, the proof is immediate. To treat the general case, use Egorov's theorem (Prob. 3, p. 131).

6. A measurable function $g(y)$ defined on a set Y with a measure λ is said to be *equimeasurable* with a measurable function $f(x)$ defined on a set X with a measure μ if $\lambda\{y: g(y) \leqslant C\} = \mu\{x: f(x) \leqslant C\}$ for arbitrary C. Prove that given any measurable function $f(x)$ on an abstract set X with measure μ, there exists a nondecreasing equimeasurable function $g(y)$ on the interval $[0, 1]$ with ordinary Lebesgue measure.

Hint. If $F(\alpha) = \mu\{x: f(x) \leqslant \alpha\}$, then

$$g(y) = \inf_{F(\alpha) \leqslant y} \alpha.$$

7. A function $f(x)$ defined on a block $\mathbf{B}$ is called a *Borel function* if every set $E(f; c) = \{x: f(x) < c\}$ is a Borel set. Show that every measurable function $f(x)$ on the block $\mathbf{B}$ can be made into a Borel function by suitable modification on a set of measure zero.

Hint. Each of the countably many sets

$$E\left(f; \frac{k+1}{2^n}\right) - E\left(f; \frac{k}{2^n}\right)$$

becomes a Borel set after deleting some set of measure zero. The union of all the deleted sets is a set of measure zero, on which we can set $f(x) \equiv 0$, say.

8

AXIOMATIC MEASURE THEORY

We now consider another way of constructing a theory of the integral, whose starting point is a family of subsets of an arbitrary set X equipped with a countably additive measure. The new approach is both simple and straightforward, if we use our prior knowledge of the Daniell scheme.

8.1. Elementary, Borel and Lebesgue Measures

A family $\mathfrak{A}$ of subsets of a set X is called a *ring* if it has the following two properties:

1) If $A \in \mathfrak{A}$, $B \in \mathfrak{A}$, then $A \cup B \in \mathfrak{A}$, $AB \in \mathfrak{A}$.
2) If $A \in \mathfrak{A}$, $B \in \mathfrak{A}$ and $B \subset A$, then $A - B \in \mathfrak{A}$.

The sets belonging to a ring $\mathfrak{A}$ are called *elementary sets*. The set X itself may or may not belong to $\mathfrak{A}$. In the latter case, it is assumed that X is the union of a countable number of elementary sets.

By a *countably additive measure* we mean a finite nonnegative additive set function $\mu(A)$, which is defined on a ring $\mathfrak{A}$ and has the following property:

a) If $A_1, \ldots, A_n, \ldots$ is any sequence of disjoint sets belonging to $\mathfrak{A}$ whose union

$$A = \bigcup_{n=1}^{\infty} A_n$$

also belongs to $\mathfrak{A}$, then

$$\mu(A) = \mu(A_1) + \cdots + \mu(A_n) + \cdots.$$

For brevity, such a measure is also called an *elementary measure*.

Two further properties of an elementary measure are easily deduced from Property a:

b) If $A_1 \subset A_2 \subset \cdots$ is any increasing sequence of sets belonging to $\mathfrak{A}$, whose union

$$A = \bigcup_{n=1}^{\infty} A_n$$

also belongs to $\mathfrak{A}$, then

$$\mu(A) = \lim_{n \to \infty} \mu(A_n).$$

In fact,

$$A = A_1 \cup (A_2 - A_1) \cup \cdots,$$

where the sets $A_1, A_2 - A_1, \ldots$ are disjoint, and hence, according to Property a,

$$\mu(A) = \mu(A_1) + \mu(A_2 - A_1) + \cdots = \lim_{n \to \infty} \mu(A_n).$$

c) If $A_1 \supset A_2 \supset \cdots$ is any decreasing sequence of sets belonging to $\mathfrak{A}$, whose intersection

$$A = \bigcap_{n=1}^{\infty} A_n$$

also belongs to $\mathfrak{A}$, then

$$\mu(A) = \lim_{n \to \infty} \mu(A_n).$$

This follows from the preceding property by taking complements (relative to A_1).

A ring $\mathfrak{A}$ is called a *σ-ring* if given any sequence $A_1, \ldots, A_n, \ldots$ of disjoint sets belonging to $\mathfrak{A}$ and contained in a fixed (but arbitrary) set $A_0 \in \mathfrak{A}$, their union

$$A = \bigcup_{n=1}^{\infty} A_n$$

also belongs to $\mathfrak{A}$. The restriction to disjoint sets can be dropped immediately. In fact, if the sets A_n are allowed to intersect, we can write

$$A = A_1 \cup (A_2 - A_1A_2) \cup (A_3 - A_1A_3 - A_2A_3) \cup \cdots,$$

where the sets on the right are now disjoint, and moreover belong to $\mathfrak{A}$ and are contained in A_0. Similarly, the intersection of any sequence of sets $A_1, \ldots, A_n, \ldots$ belonging to $\mathfrak{A}$ also belongs to $\mathfrak{A}$. To see this, we merely observe that

$$A_1 - \bigcap_{n=1}^{\infty} A_nA_1 = \bigcap_{n=1}^{\infty} (A_1 - A_nA_1)$$

belongs to $\mathfrak{A}$, and hence so does

$$\bigcap_{n=1}^{\infty} A_n = A_1 - \left(A_1 - \bigcap_{n=1}^{\infty} A_n A_1\right).$$

The sets belonging to a σ-ring are called (*abstract*) *Borel sets*.

A countably additive finite measure μ defined on a σ-ring is called a *Borel measure*, and relative to the measure μ, the sets $A \in \mathfrak{A}$ are called *measurable* (more exactly, μ-*measurable*). A σ-ring $\mathfrak{A}$ with a Borel measure μ is called a σ_μ-*ring* if every subset of a set $Z \in \mathfrak{A}$ of measure zero also belongs to $\mathfrak{A}$ (and hence has measure zero itself). In this case, the measure μ is called a *finitely-Lebesgue measure*.

A ring $\mathfrak{A}$ of subsets of the set X is called a Σ-*ring* if given any sequence $A_1, \ldots, A_n, \ldots$ of disjoint sets belonging to $\mathfrak{A}$, their union also belongs to $\mathfrak{A}$. Just as before, it is easy to see that the sets need not be disjoint, and that the intersection

$$\bigcap_{n=1}^{\infty} A_n$$

belongs to $\mathfrak{A}$. It is always assumed that the set X itself is an element of the Σ-ring. The sets belonging to a Σ-ring are called *generalized Borel sets*.

A nonnegative additive set function $\mu(A)$, defined on a Σ-ring $\mathfrak{A}$ and taking ∞ as a possible value, is called a *generalized Borel measure* if

$$\mu\left(\bigcup_{n=1}^{\infty} A_n\right) = \mu(A_1) + \cdots + \mu(A_n) + \cdots,$$

where $A_1, \ldots, A_n, \ldots$ is any sequence of disjoint sets belonging to $\mathfrak{A}$ (here the left and right-hand sides are allowed to take the value ∞). Relative to the measure μ, the sets $A \in \mathfrak{A}$ are called *measurable* (more exactly, μ-*measurable*), and the sets such that $\mu(A) < \infty$ are called *summable* (μ-*summable*). If $\mu(X) = \infty$, it is assumed that $X = X_1 \cup X_2 \cup \cdots$, where $X_1 \subset X_2 \subset \cdots$ and every X_n ($\in \mathfrak{A}$) is of finite measure. A Σ-ring $\mathfrak{A}$ with a generalized Borel measure μ is called a Σ_μ-*ring* if every subset of a set $Z \in \mathfrak{A}$ of measure zero also belongs to $\mathfrak{A}$. In this case, the measure μ is called a *Lebesgue measure*.

Example 1. As in Chap. 2, let X be the domain of a space of elementary functions, equipped with an elementary integral, and use the Daniell scheme to construct an integral. Then the "Daniell-measurable" sets (see Chap. 7) form a Σ_μ-ring, on which μ is a Lebesgue measure, while the "Daniell-summable" sets form a σ_μ-ring, on which μ is a finitely-Lebesgue measure.

Example 2. The bounded classical Borel sets (see Sec. 7.8) on the real line $-\infty < x < \infty$ form a σ-ring, on which ordinary Lebesgue measure (or a Lebesgue-Stieltjes measure) is a Borel measure (but not a finitely-Lebesgue measure). The family of all Borel sets on the real line forms a Σ-ring, on which ordinary Lebesgue measure is a generalized Borel measure.

Example 3. The half-open intervals $(\alpha, \beta]$ and their finite combinations form a ring (but not a σ-ring), on which ordinary Lebesgue measure (or a Lebesgue-Stieltjes measure) is an elementary measure.

8.2. Lebesgue and Borel Extensions of an Elementary Measure

We begin by proving

THEOREM 1. *Let μ be an elementary measure defined on a ring $\mathfrak{A}$ of subsets of a set X, and let H be the family of finite linear combinations*

$$h(x) = \sum_{k=1}^{m} \alpha_k \chi_{E_k}(x) \tag{1}$$

of characteristic functions of the elementary sets. Then H has all the properties of a family of elementary functions, and the integral

$$Ih = \int_X h(x)\mu(dx) = \sum_{k=1}^{m} \alpha_k \mu(E_k) \tag{2}$$

has all the properties of an elementary integral.

Proof. Obviously H is a linear space, and hence satisfies Axiom a, p. 23. Moreover, the sets E_k figuring in (1) can always be chosen to be disjoint, and then

$$|h(x)| = \sum_{k=1}^{m} |\alpha_k| \, \chi_{E_k}(x)$$

also belongs to H, which proves Axiom b, p. 23, while

$$\min(h, 1) = \sum_{k=1}^{m} \min(\alpha_k, 1)\chi_{E_k}(x) \in H,$$

which verifies Stone's Axiom c, p. 119. As for Stone's Axiom d, if $\mu(X) < \infty$, then $h_n(x) \equiv 1$ is a sequence of nonnegative functions such that

$$Ih_n > 0, \qquad \sup_n h_n(x) > 0 \qquad (x \in X), \tag{3}$$

while if $\mu(X) = \infty$, then, by hypothesis, X is the union of an increasing sequence of elementary sets X_n and the sequence of nonnegative functions $h_n(x) = \chi_{X_n}(x)$ again satisfies (3). Thus H has all the properties of a space of elementary functions.

Next we turn to the proposed elementary integral (2). The fact that I is linear and nonnegative, i.e., that I satisfies Axioms 1 and 2, p. 24, is immediately apparent. As for Axiom 3, we must show that if $h_n(x) \searrow 0$ for every x, then $Ih_n \to 0$. First we note that if the integral of an elementary function over an elementary subset $A \subset X$ is defined by the natural formula $I_A h = I_A \chi_A h$ (as in Sec. 6.7), then $|I_A h| \leqslant M\mu(A)$ if $|h(x)| \leqslant M$, and $I_{A \cup B} h = I_A h + I_B h$ if A and B are disjoint. Now let $E \subset X$ be the set where $h_1(x) > 0$ [there is no need to consider the set where $h_1(x) = 0$], and let

$$A_n^{(m)} = E \cap \{x: h_n(x) \leqslant 1/m\},$$

Since

$$h_n(x) = \sum_{k=1}^{r_n} \alpha_k^{(n)} \chi_{E_k^{(n)}}(x),$$

where the $E_k^{(n)}$ can be regarded as disjoint (for fixed n), the set $A_n^{(m)}$ is a finite union of certain $E_k^{(n)}$ (in fact, those for which $\alpha_k^{(n)} \leqslant 1/m$), and hence is an elementary set. For fixed m, the sets $A_n^{(m)}$ enlarge as n increases, and clearly

$$E = \bigcup_{n=1}^{\infty} A_n^{(m)} \qquad (m = 1, 2, \ldots).$$

Therefore

$$\mu(E) = \lim_{n \to \infty} \mu(A_n^{(m)}),$$

by Property b, p. 151. In other words, we can find an integer $n = n(m)$ such that

$$\mu(A_{n(m)}^{(m)}) > \mu(E) - \frac{1}{m},$$

and hence

$$\mu(E - A_{n(m)}^{(m)}) < \frac{1}{m}.$$

It follows that

$$Ih_n = I_{E - A_{n(m)}^{(m)}} h_n + I_{A_{n(m)}^{(m)}} h_n < \frac{1}{m} M + \frac{1}{m} \mu(E) = \frac{1}{m} [M + \mu(E)]$$

if $n > n(m)$, where $M = \max h_1(x)$. But then $Ih_n \to 0$, since m can be made arbitrarily large. Thus, finally, (2) has all the properties of an elementary integral, and the theorem is proved.

Next we show that every elementary measure can be extended to a Lebesgue measure:

THEOREM 2. *Let $\mathfrak{A}$ be a ring, equipped with an elementary measure μ. Then there is a Σ_μ-ring $\overline{\mathfrak{A}}$, equipped with a Lebesgue measure $\overline{\mu}$, such that $\overline{\mathfrak{A}} \supset \mathfrak{A}$ and $\overline{\mu}(A) = \mu(A)$ for every $A \in \mathfrak{A}$.*

Proof. Let H denote the set of linear combinations (1) of characteristic functions of sets of $\mathfrak{A}$, and use formula (2) to define an elementary integral on H. Then Theorem 1 allows us to apply all the results of Chaps. 2 and 6, thereby constructing a space $L(X)$ of summable functions and a Σ_μ-ring $\overline{\mathfrak{A}}$ of measurable sets equipped with a measure $\overline{\mu}$, where $\overline{\mu}$ is a Lebesgue measure (as already noted in Example 1, p. 152). Obviously, $\overline{\mathfrak{A}} \supset \mathfrak{A}$ and $\overline{\mu}(A) = I\chi_A = \mu(A)$ for every $A \in \mathfrak{A}$, and hence the measure $\overline{\mu}$ can serve as the desired Lebesgue extension of the elementary measure μ.

Remark. It is clear from the very construction of $L(X)$ that the original ring $\mathfrak{A}$ is *sufficient* (see Sec. 7.3), i.e., linear combinations of characteristic functions of sets of $\mathfrak{A}$ are dense in $L(X)$. The ring $\mathfrak{A}$ is also *completely sufficient*, as defined in Sec. 7.4. In fact, let Z be a set of $\overline{\mu}$-measure zero. Then, given any $\varepsilon > 0$, there exists a nondecreasing sequence of nonnegative functions $h_n \in H$ such that $Ih_n < \varepsilon$ and $\sup_n h_n(x) \geqslant 1$ on Z. Suppose that

$$h_n(x) = \sum_{k=1}^{m} \alpha_{kn}\chi_{E_{kn}}(x),$$

where the μ-summable sets E_{kn} are disjoint (for fixed n). Let G_n be the union of the sets E_{kn} with coefficients $\alpha_{kn} \geqslant \frac{1}{2}$. Then $\mu(G_n) < 2\varepsilon$, since $Ih_n < \varepsilon$. Moreover

$$G = \bigcup_{n=1}^{\infty} G_n \supset Z,$$

since $\sup h_n(x) \geqslant 1$ on Z. But the sequence h_n is nondecreasing, and hence $G_1 \subset G_2 \subset \cdots$. Therefore $\mu(G) = \lim \mu(G_n) \leqslant 2\varepsilon$, as required (note that $G \in \mathfrak{A}_\sigma$).

THEOREM 3. *Let $\mathfrak{A}$ be a $\sum_\mu$-ring of subsets of X, equipped with a Lebesgue measure μ such that $\mu(X) < \infty$, and let $\overline{\mathfrak{A}}$ and $\overline{\mu}$ be constructed as in Theorem 2. Then $\overline{\mathfrak{A}} = \mathfrak{A}$ and $\overline{\mu}(A) = \mu(A)$ for every $A \in \mathfrak{A}$.*

Proof. Roughly speaking, the construction of Theorem 2 leads to nothing new if the original ring is a $\sum_\mu$-ring and the elementary measure is a Lebesgue measure. First let Z be a set of $\overline{\mu}$-measure zero. As already noted, the elementary sets form a completely sufficient ring. Therefore, given any $m = 1, 2, \ldots$, the set Z can be covered by a finite or countable union E_m of elementary sets E_{mk} $(k = 1, 2, \ldots)$ such that $\mu(E_m) < 1/m$. Clearly $E_m \in \mathfrak{A}$, and moreover $Z \subset \bigcap E_m = E$, where $E \in \mathfrak{A}$, $\mu(E) = 0$. But then $Z \in \mathfrak{A}$, $\mu(Z) = 0$, since μ is a Lebesgue measure.

Next let A be an arbitrary $\overline{\mu}$-summable set. Then, according to Theorem 3, p. 140, A can be represented in the form $A = G - Z$, where $G \in \mathfrak{A}_{\sigma\delta}$, $\overline{\mu}(Z) = 0$. Therefore $G \in \mathfrak{A}$, since $\mathfrak{A}$ is a $\sum_\mu$-ring, and moreover $Z \in \mathfrak{A}$, as just proved. It follows that $A = G - Z \in \mathfrak{A}$, and hence $\overline{\mathfrak{A}} = \mathfrak{A}$, as asserted, where obviously $\overline{\mu}(A) = \mu(A)$.

Remark. Suppose we start from a σ_μ-ring $\mathfrak{A}$ of subsets of X, where $X \notin \mathfrak{A}$, equipped with a *finitely-Lebesgue* measure μ. Then in general $\overline{\mathfrak{A}}$ is larger than $\mathfrak{A}$ not only because it contains sets of infinite $\overline{\mu}$-measure, but also because it contains sets of finite $\overline{\mu}$-measure not contained in $\mathfrak{A}$ itself. For example, let $\mathfrak{A}$ be the σ_μ-ring of bounded Lebesgue-measurable sets on the real line $-\infty < x < \infty$, equipped with ordinary Lebesgue measure. Then $\overline{\mathfrak{A}}$ consists of all Lebesgue-measurable sets on $-\infty < x < \infty$, including unbounded summable sets.

In general, a given ring $\mathfrak{A}$ equipped with an elementary measure μ can be extended in many ways to a Σ_μ-ring equipped with a Lebesgue measure $\overline{\mu}$ such that $\overline{\mu}(A) = \mu(A)$ if $A \in \mathfrak{A}$. Every such measure $\overline{\mu}$ will be called a *Lebesgue extension* of the elementary measure μ. Two different Lebesgue extensions of μ always lead to another, as shown by

THEOREM 4. *Given an elementary measure μ, defined on a ring $\mathfrak{A}$ of subsets of a set X, let μ_1 and μ_2 be two Lebesgue extensions of μ, defined on Σ_μ-rings $\mathfrak{A}_1$ and $\mathfrak{A}_2$ respectively. Let $\mathfrak{B}$ be the family of all subsets $A \subset X$ on which both μ_1 and μ_2 are defined and $\mu_1(A) = \mu_2(A)$. Then $\mathfrak{B}$ is a Σ_μ-ring, and the set function*

$$\nu(A) = \mu_1(A) = \mu_2(A),$$

called the intersection of the measures μ_1 and μ_2, is a Lebesgue measure on $\mathfrak{B}$ (in fact, a Lebesgue extension of μ).

Proof. If $A, B \in \mathfrak{B}$ and if $A \subset B$, then $B - A \in \mathfrak{B}$, since $B - A \in \mathfrak{A}_1$, $B - A \in \mathfrak{A}_2$ and

$$\mu_1(B - A) = \mu_1(B) - \mu_1(A) = \mu_2(B) - \mu_2(A) = \mu_2(B - A).$$

Moreover, if $A_1, \ldots, A_n, \ldots$ is a sequence of disjoint sets of $\mathfrak{B}$, with union

$$A = \bigcup_{n=1}^{\infty} A_n,$$

then A also belongs to $\mathfrak{B}$, since $A \in \mathfrak{A}_1$, $A \in \mathfrak{A}_2$ and

$$\mu_1(A) = \sum_{n=1}^{\infty} \mu_1(A_n) = \sum_{n=1}^{\infty} \mu_2(A_n) = \mu_2(A).$$

Finally, if $E \subset E_0 \in \mathfrak{B}$ and if $\mu_1(E_0) = \mu_2(E_0) = 0$, then E belongs to both of the Σ_μ-rings $\mathfrak{A}_1$ and $\mathfrak{A}_2$, in each of which it has Lebesgue measure zero, i.e., E_0 belongs to $\mathfrak{B}$ and $\nu(E_0) = 0$. It follows that $\mathfrak{B}$ is a Σ_μ-ring (containing $\mathfrak{A}$) and that ν is a Lebesgue measure on $\mathfrak{B}$. The fact that $\nu(A) = \mu(A)$ if $A \in \mathfrak{A}$ is immediately apparent.

Similarly, we can show that the intersection of an *arbitrary* number of Lebesgue extensions of an elementary measure μ (defined in the obvious way) is itself a Lebesgue extension of μ. In particular, let μ^* be the intersection of *all* Lebesgue extensions of the measure μ. Then clearly μ^* is the *smallest* Lebesgue extension of μ, in the sense that the intersection of μ^*

with any other Lebesgue extension of μ is again μ^*. In fact, as we now show, μ^* has already been constructed in the proof of Theorem 2:

THEOREM 5. *The Lebesgue extensions $\bar{\mu}$ and μ^* of the elementary measure μ coincide.*

Proof. We need only show that the intersection μ_1 of any Lebesgue extension of μ with the Lebesgue extension $\bar{\mu}$ coincides with $\bar{\mu}$. Let $\mathfrak{A}$ be the ring on which μ is defined, and let $\mathfrak{A}_1$, $\bar{\mathfrak{A}}$ be the Σ_μ-rings on which μ_1, $\bar{\mu}$ are defined. Then it is enough to prove that $\mathfrak{A}_1 = \bar{\mathfrak{A}}$. Obviously $\mathfrak{A} \subset \mathfrak{A}_1 \subset \bar{\mathfrak{A}}$, and hence, by the considerations of Sec. 7.2, $\bar{\mathfrak{A}} \subset \bar{\mathfrak{A}}_1 \subset \bar{\bar{\mathfrak{A}}}$ if we extend the measures μ, μ_1 and $\bar{\mu}$ themselves. But according to Theorem 3, $\bar{\mathfrak{A}}_1 = \mathfrak{A}_1$, $\bar{\bar{\mathfrak{A}}} = \bar{\mathfrak{A}}$, since $\mathfrak{A}_1$ and $\bar{\mathfrak{A}}$ are already Σ_μ-rings. It follows that $\bar{\mathfrak{A}} \subset \mathfrak{A}_1 \subset \bar{\mathfrak{A}}$, i.e., $\mathfrak{A}_1 = \bar{\mathfrak{A}}$, and the theorem is proved.

Remark. It can be shown that given any $\bar{\mu}$-nonmeasurable set $Y \subset X$, there is always a Lebesgue extension of the measure $\bar{\mu}$ in which Y is measurable (see Prob. 2, p. 178).

Next we study *Borel extensions* of the elementary measure μ. Instead of the Σ_μ-ring $\bar{\mathfrak{A}}$ of $\bar{\mu}$-measurable sets constructed in Theorem 2, consider the σ_μ-ring of $\bar{\mu}$-summable sets, which we continue to denote by the symbol $\bar{\mathfrak{A}}$. Then $\bar{\mu}$ is a Borel extension of the elementary measure μ, in the sense that $\bar{\mathfrak{A}}$ is a σ-ring (in fact, a σ_μ-ring) containing $\mathfrak{A}$ and $\bar{\mu}(A) = \mu(A)$ if $A \in \mathfrak{A}$.[1] However, in general $\bar{\mu}$ is not the "smallest Borel extension" of μ, which is constructed as follows: Let $\mathfrak{A}^*$ be the intersection of all σ-rings containing $\mathfrak{A}$, equipped with the Borel measure μ^* defined by the formula $\mu^*(A) = \bar{\mu}(A)$ if $A \in \mathfrak{A}^*$ (note that $\mathfrak{A}^* \subset \bar{\mathfrak{A}}$). Clearly, $\mathfrak{A}^*$ is the smallest σ-ring containing $\mathfrak{A}$, a fact which suggests calling μ^* the *smallest Borel extension* of μ. To justify this definition, we must prove that it is consistent with our previous definition of the smallest (Lebesgue) extension of μ:

THEOREM 6. *The intersection μ_1 of any Borel extension of μ with the Borel extension μ^* coincides with μ^*.*

Proof. First we note that Theorem 4 obviously remains true if we change the word "Lebesgue" to "Borel" and the symbol Σ_μ to σ_μ. Let $\mathfrak{A}$ be the ring on which μ is defined, and let $\mathfrak{A}_1$, $\mathfrak{A}^*$ be the σ-rings on which μ_1, μ^* are defined. Then it is enough to prove that $\mathfrak{A}_1 = \mathfrak{A}^*$. But $\mathfrak{A}_1 \subset \mathfrak{A}^*$, since μ_1 is an intersection with the measure μ^*, while on the other hand $\mathfrak{A}^* \subset \mathfrak{A}_1$, since $\mathfrak{A}_1$ is a σ-ring containing $\mathfrak{A}$ and $\mathfrak{A}^*$ is the smallest such σ-ring. Therefore $\mathfrak{A}_1 = \mathfrak{A}^*$, as required.

---

[1] In fact, $\bar{\mu}$ is a "finitely-Lebesgue extension" of μ, in an obvious sense.

Remark. As already noted, $\bar{\mu}$ and μ^* do not coincide in general (unlike the case of Lebesgue extensions). In fact, $\bar{\mathfrak{A}}$ is obtained from $\mathfrak{A}^*$ by forming all possible unions of sets $A \in \mathfrak{A}^*$ with subsets of sets $Z \in \mathfrak{A}^*$ of measure zero. To see this, we note that on the one hand, the smallest finitely-Lebesgue extension of μ must contain all the sets so obtained, while on the other hand, the construction clearly leads to a finitely-Lebesgue extension of μ.

8.3. Construction of the Integral from a Lebesgue Measure

Let $\mathfrak{A}$ be a Σ_μ-ring of subsets of a given set X, equipped with a Lebesgue measure μ. Then to construct a theory of Lebesgue integration on X, we proceed as follows: A function $\varphi(x)$, defined on X, is said to be *μ-measurable* if the set

$$E(\varphi; c) = \{x: \varphi(x) > c\}$$

is μ-measurable for arbitrary real c. If $\varphi(x)$ is μ-measurable, the set

$$\{x: c < \varphi(x) \leqslant d\} = E(\varphi; c) - E(\varphi; d)$$

is also μ-measurable for arbitrary c and d $(c < d)$. The *Lebesgue integral* of a nonnegative measurable function $\varphi(x)$ is defined by the formula

$$I\varphi = \lim_{\varepsilon \to 0} \sum_{n=1}^{\infty} n\varepsilon\mu\{x: n\varepsilon < \varphi(x) \leqslant (n+1)\varepsilon\}, \tag{4}$$

provided the limit exists, and the function $\varphi(x)$ is then said to be *μ-summable.* A μ-measurable function $\varphi(x)$ of variable sign is said to be μ-summable if its positive and negative parts $\varphi^+(x)$ and $\varphi^-(x)$ are μ-summable, and the Lebesgue integral of φ is then defined as the difference $I\varphi^+ - I\varphi^-$.

Next we show that the integral I defined by (4) has all the customary properties of the integral, as contained, say, in the theorems of Chap. 2, Secs. 6–9. It would be tedious to verify this directly, but there is no need to do so. In fact, starting from the Lebesgue measure μ, we construct the family H of linear combinations of characteristic functions of μ-summable sets, equipped with the integral (2) of Theorem 1, afterwards using this theorem and the Daniell scheme to extend I to a space L. Then, just as in Chap. 6, we consider the resulting "Daniell-measurable" sets and functions. According to Theorem 3, p. 155, the "Daniell-measurable" sets are just the original μ-measurable sets, while, according to Theorem 5, p. 120, the "Daniell-measurable" functions are just the μ-measurable functions defined above. Moreover, according to Sec. 6.6, the "Daniell-summable" functions are just the μ-summable functions defined above, and the "Lebesgue-Daniell" integral of a given μ-summable function φ has the same value as the Lebesgue integral (4). Thus all the properties of the integral established in Chap. 2, remain valid for the Lebesgue integral with the present definition.

8.4. Signed Borel Measures

So far, the measure μ defined on a ring $\mathfrak{A}$ of subsets of a set X has been assumed to be nonnegative. We now consider "signed measures," i.e., measures which can take values of either sign, eventually proving that every such measure is the difference between two nonnegative measures (see Theorem 8 below).

Let $\mu(E)$ be a finite additive set function defined on a σ-ring $\mathfrak{A}$, and suppose that $\mu(E)$ can take values of either sign. Suppose further that $\mu(E)$ is countably additive in the sense that

$$\mu(E) = \sum_{n=1}^{\infty} \mu(E_n)$$

if $E_1, \ldots, E_n, \ldots$ is any sequence of disjoint sets of $\mathfrak{A}$, all contained in some set $A_0 \in \mathfrak{A}$. Then $\mu(E)$ is called a *(signed) Borel measure*. Obviously, $\mu(E)$ still satisfies Properties b and c, p. 151.

THEOREM 7. *Let $\mathfrak{A}$ be a σ-ring equipped with a signed Borel measure μ. Then the set function*

$$\lambda(E) = \sup_{A \subset E} \mu(A), \tag{5}$$

where the least upper bound is taken with respect to all μ-measurable subsets of E, defines a nonnegative Borel measure on $\mathfrak{A}$.

Proof. After first observing that $\lambda(E) \geqslant 0$ and $\lambda(E) \geqslant \mu(E)$ [since A can always be chosen to be either the empty set or the set E itself] and that the possibility $\lambda(E) = +\infty$ is not excluded *a priori*, we proceed to establish the proof in three steps:

Step 1. $\lambda(E)$ is subadditive, i.e., if $E_1, E_2, \ldots$ is a sequence of disjoint sets of $\mathfrak{A}$ (all contained in some set $A_0 \in \mathfrak{A}$), then[2]

$$\lambda(E_1 \cup E_2 \cup \cdots) \leqslant \lambda(E_1) + \lambda(E_2) + \cdots. \tag{6}$$

In fact, let A be any μ-measurable set contained in $E_1 \cup E_2 \cup \cdots$. Then $A = AE_1 \cup AE_2 \cup \cdots$ represents A as a union of disjoint sets. Therefore, since μ is countably additive,

$$\mu(A) = \mu(AE_1) + \mu(AE_2) + \cdots \leqslant \lambda(AE_1) + \lambda(AE_2) + \cdots,$$

and taking the least upper bound of the left-hand side with respect to A, we obtain (6).

[2] Clearly, $E_1 \cup E_2 \cup \cdots$ is μ-measurable, since $\mathfrak{A}$ is a σ-ring.

Step 2. $\lambda(E)$ is finite. Suppose, to the contrary, that $\lambda(E) = \infty$ for some μ-measurable set E. Then, by induction, we can construct a sequence of μ-measurable sets

$$E_0 \supset E_1 \supset \cdots \supset E_m \supset \cdots \tag{7}$$

such that

$$\lambda(E_m) = \infty, \qquad |\mu(E_m)| \geqslant m. \tag{8}$$

First set $E_0 = E$, a choice which obviously satisfies (8) for $m = 0$. Then suppose sets $E_0 \supset E_1 \supset \cdots \supset E_{m-1}$ satisfying (8) have already been constructed. Since $\lambda(E_{m-1}) = \infty$, there is a μ-measurable set $A_m \subset E_{m-1}$ such that

$$\mu(A_m) \geqslant m + |\mu(E_{m-1})|.$$

If $\lambda(A_m) = \infty$, we can set $E_m = A_m$, thereby completing the induction. However, if $\lambda(A_m)$ is finite, $\lambda(E_{m-1} - A_m)$ must be infinite, since otherwise (6) would be contradicted, and moreover

$$|\mu(E_{m-1} - A_m)| \geqslant \mu(A_m) - |\mu(E_{m-1})| \geqslant m.$$

Thus, in this case, we complete the induction by choosing $E_m = E_{m-1} - A_m$. In any event, once having constructed a sequence of sets (7) satisfying (8), we note that the numerical sequence $\mu(E_m)$ must have a limit [equal to $\mu(\bigcap E_m)$], since the measure μ is countably additive. But this contradicts (8), and hence $\lambda(E)$ is finite, as asserted.

Step 3. $\lambda(E)$ is countably additive. Let $E_1, E_2, \ldots$ be a sequence of disjoint sets of $\mathfrak{A}$, all contained in some set $A_0 \in \mathfrak{A}$. Given any $\varepsilon > 0$ and any integer $m = 1, 2, \ldots$, let $A_m \subset E_m$ be a set such that

$$\lambda(E_m) < \mu(A_m) + \frac{\varepsilon}{2^m}.$$

Such a set A_m exists, since, as just shown, $\lambda(E_m)$ is finite. Then

$$\begin{aligned}\lambda(E_1) + \lambda(E_2) + \cdots &< \mu(A_1) + \mu(A_2) + \cdots + \varepsilon \\ &= \mu(A_1 \cup A_2 \cup \cdots) + \varepsilon \leqslant \lambda(E_1 \cup E_2 \cup \cdots) + \varepsilon,\end{aligned}$$

and making ε approach zero, we obtain

$$\lambda(E_1) + \lambda(E_2) + \cdots \leqslant \lambda(E_1 \cup E_2 \cup \cdots). \tag{9}$$

Together, the inequalities (9) and (6) imply that $\lambda(E)$ is countably additive. This completes the proof.

THEOREM 8. *A signed Borel measure $\mu(E)$ can be represented as the difference between two nonnegative Borel measures.*

Proof. Let

$$\nu(E) = \lambda(E) - \mu(E),$$

where $\lambda(E)$ is the nonnegative Borel measure (5). Since $\lambda(E)$ and $\mu(E)$ are countably additive, so is $\nu(E)$. Moreover, $\nu(E)$ is nonnegative, since $\lambda(E) \geqslant \mu(E)$. Therefore $\nu(E)$ is a nonnegative Borel measure, and

$$\mu(E) = \lambda(E) - \nu(E) \tag{10}$$

is a representation of the desired type.

Remark. The representation (10) is not unique. In fact, let $\tau(E)$ be any nonnegative Borel measure defined on $\mathfrak{A}$. Then, besides (10), we can write

$$\mu(E) = [\lambda(E) + \tau(E)] - [\nu(E) + \tau(E)] = \lambda_1(E) - \nu_1(E). \tag{11}$$

Moreover, by an obvious modification of the argument given in Sec. 4.7.2, (11) is the most general representation of $\mu(E)$ as a difference between two nonnegative measures, and the measures $\lambda(E)$ and $\nu(E)$ are the smallest possible among all that can figure in (11) [hence (10) is called the *canonical representation* of μ, as on p. 83]. In particular, we have the formula

$$\nu(E) = \sup_{A \subset E} [-\mu(A)] \tag{12}$$

(where A is μ-measurable), since ν plays the same role in the representation $-\mu = \nu - \lambda$ as λ plays in the representation $\mu = \lambda - \nu$.

The nonnegative measures λ, ν and $\rho = \lambda + \nu$ are called the *positive variation*, the *negative variation* and the *total variation* of μ, respectively (cf. Sec. 4.7.3). Let $\bar{\lambda}$, $\bar{\nu}$ and $\bar{\rho}$ denote the Lebesgue extensions of λ, ν and ρ, constructed as in Sec. 8.2. Then every $\bar{\rho}$-summable set E is also $\bar{\lambda}$-summable and $\bar{\nu}$-summable, by substantially the same argument as given in Sec. 5.3. This allows us to extend the measure μ itself onto the family of all $\bar{\rho}$-summable sets, by writing

$$\bar{\mu}(E) = \bar{\lambda}(E) - \bar{\nu}(E).$$

In the terminology of Secs. 8.1 and 8.2, $\bar{\mu}$ is a finitely-Lebesgue extension of the Borel measure μ. In general, the need to avoid indeterminacies of the form $\infty - \infty$ prevents $\bar{\mu}$ from being a Lebesgue extension.

Next we define integration with respect to the signed measure μ (omitting the overbar for simplicity). A function $\varphi(x)$ is said to be *μ-measurable* if it is $\bar{\rho}$-measurable, and μ-summable if it is $\bar{\rho}$-summable. According to formula (4), p. 158, the $\bar{\rho}$-integral of a $\bar{\rho}$-summable nonnegative function $\varphi(x)$ is given by

$$I_{\bar{\rho}}\varphi = \lim_{\varepsilon \to 0} \sum_{n=1}^{\infty} n\varepsilon\bar{\rho}\{x: n\varepsilon < \varphi(x) \leqslant (n+1)\varepsilon\}.$$

The μ-integral of the same function φ is defined in the natural way:

$$I_{\mu}\varphi = \lim_{\varepsilon \to 0} \sum_{n=1}^{\infty} n\varepsilon\mu\{x: n\varepsilon < \varphi(x) \leqslant (n+1)\varepsilon\} = I_{\bar{\lambda}}\varphi - I_{\bar{\nu}}\varphi.$$

Then, for a function of variable sign, we write

$$I_\mu \varphi = I_\mu \varphi^+ - I_\mu \varphi^-,$$

as usual. It is obvious that the integrals $I_{\bar{\rho}}$ and I_μ have all the usual properties of the integral (apart from the fact that μ is signed).

8.5. Quasi-Volumes and Measure Theory

Let σ be a nonnegative quasi-volume defined on a dense set Q of subblocks B of a finite basic block $\mathbf{B}$ in Euclidean n-space. Since the set function σ is additive on the semiring Q, it is natural to ask whether σ can be extended to a (countably additive) Borel measure on some σ-ring.[3] The theorems of Sec. 8.2 are not immediately applicable, since σ is originally defined on a semiring instead of a ring and is in general not countably additive. However, as shown in Sec. 5.8, if the quasi volume σ is *continuous*, there exists a space L_σ of σ-summable functions containing the characteristic functions of *all* blocks $B \subset \mathbf{B}$ such that $I_\sigma \chi_B = \sigma(B)$. Let $\mathfrak{A}$ be the σ-ring of all σ-summable subsets $E \subset \mathbf{B}$, equipped with the measure $\mu(E) = I_\sigma \chi_E$. Then μ is the desired extension of the quasi-volume σ. Note that $\mathfrak{A}$ contains all the classical Borel sets in $\mathbf{B}$, and is actually a σ_μ-ring. The restriction to nonnegative quasi-volumes can be easily removed. In fact, a continuous quasi-volume of bounded variation can always be represented as the difference between two nonnegative quasi-volumes p and q, which are themselves continuous (see p. 100). Thus every continuous quasi-volume σ of bounded variation can be extended to a signed Borel measure μ.

Conversely, let $\mu(E)$ be a signed Borel measure defined on the σ-ring of Borel subsets of an n-dimensional basic block $\mathbf{B}$. Then, considered only on the blocks $B \subset \mathbf{B}$, the measure μ is clearly a continuous quasi-volume of bounded variation (see Theorem 2, p. 101), which, just as in Sec. 4.7.3, can be decomposed into positive and negative variations

$$p(B) = \sup \sum_{j=1}^{m} \mu(B_j), \qquad q(B) = \sup\left\{-\sum_{j=1}^{m} \mu(B_j)\right\}, \tag{13}$$

where the least upper bounds are taken with respect to all sets of disjoint subblocks $B_j \subset B$ $(j = 1, \ldots, m)$. On the other hand, according to formulas (5) and (12), μ gives rise to a pair of nonnegative Borel measures

$$\lambda(E) = \sup_{A \subset E} \mu(A), \qquad \nu(E) = \sup_{A \subset E} [-\mu(A)], \tag{14}$$

[3] The symbol σ is used here in two different senses, but the context precludes any possibility of confusion. The present discussion is closely related to that at the beginning of Sec. 7.6.

where the least upper bounds are taken with respect to all μ-measurable subsets of E. To complete the correspondence between signed measures and quasi-volumes, we must still prove the consistency of the formulas (13) and (14):

THEOREM 9. *The relations*

$$\lambda(B) = p(B), \qquad \nu(B) = q(B)$$

hold for every block $B \subset \mathbf{B}$.

Proof. Clearly

$$\lambda(B) = \sup_{A \subset B} \mu(A) \geqslant p(B), \tag{15}$$

since the least upper bound is taken with respect to a larger family of sets than in (13). On the other hand, as shown on p. 144, given any $\varepsilon > 0$, there is a finite union of blocks $F = B_1 \cup \cdots \cup B_m$ such that

$$v(E - EF) + v(F - EF) < \varepsilon,$$

where $v(E) = p(E) + q(E)$ and E is any v-measurable set. But then

$$|\mu(E - EF)| + |\mu(F - EF)| < \varepsilon,$$

since $|\mu| \leqslant v$. Clearly, there is no loss of generality in assuming that the sets B_j are disjoint and contained in B. It follows that

$$\lambda(B) \leqslant \sup \sum_{j=1}^{m} \mu(B_j) + \varepsilon = p(B) + \varepsilon,$$

which implies

$$\lambda(B) \leqslant p(B), \tag{16}$$

since ε is arbitrary. Comparing (15) and (16), we find that $\lambda(B) = p(B)$, and hence $\nu(B) = q(B)$ also, as required.

8.6. The Hahn Decomposition

Besides the representation of a signed Borel measure μ as the difference between two nonnegative measures, there is another representation of μ involving the set X itself:

THEOREM 10 (*Hahn decomposition*). *Let $\mathfrak{A}$ be a σ-ring of subsets of X, equipped with a signed Borel measure μ. Then X is the union of two disjoint generalized Borel sets X^+ and X^- such that*[4]

$$\begin{aligned} \mu(E) &\geqslant 0 \quad \text{if} \quad E \subset X^+, \\ \mu(E) &\leqslant 0 \quad \text{if} \quad E \subset X^- \end{aligned} \tag{17}$$

for every Borel set E.

[4] Generalized Borel sets in a σ-ring $\mathfrak{A}$ are defined in the obvious way, i.e., as countable unions of sets $E_1, E_2, \ldots \in \mathfrak{A}$ not known to be contained in a fixed set $E_0 \in \mathfrak{A}$.

Proof. Assuming first that $X \in \mathfrak{A}$, for every $n = 1, 2, \ldots$, we find a set $E_n \subset X$ such that

$$\mu(E_n) > \lambda(X) - \frac{1}{2^n},$$

where λ is the positive variation of μ. We then write

$$X^+ = \bigcup_{m=1}^{\infty} \bigcap_{n=1}^{\infty} E_n, \qquad X^- = X - X^+ = \bigcap_{m=1}^{\infty} \bigcup_{n=1}^{\infty} (X - E_n),$$

i.e., a point belongs to X^+ if it belongs to all the sets E_n starting from a certain value of n, and X^- is the complement of X^+. Since

$$\lambda(E_n) \geqslant \mu(E_n) > \lambda(X) - \frac{1}{2^n},$$

$$\lambda(X - E_n) = \lambda(X) - \lambda(E_n) < \frac{1}{2^n},$$

$$\nu(E_n) = \lambda(E_n) - \mu(E_n) \leqslant \lambda(X) - \mu(E_n) < \frac{1}{2^n},$$

we have

$$\lambda(X^-) \leqslant \lambda\left(\bigcup_{n=m}^{\infty} (X - E_n)\right) \leqslant \sum_{n=m}^{\infty} \lambda(X - E_n) < \frac{1}{2^{m-1}},$$

and hence $\lambda(X^-) = 0$. On the other hand, for any m and $n \geqslant m$,

$$\nu\left(\bigcap_{n=m}^{\infty} E_n\right) \leqslant \nu(E_n),$$

and hence

$$\nu\left(\bigcap_{n=m}^{\infty} E_n\right) = 0, \qquad \nu(X^+) \leqslant \sum_{m=1}^{\infty} \nu\left(\bigcap_{n=m}^{\infty} E_n\right) = 0.$$

Therefore

$$\lambda(X^-) = 0, \qquad \nu(X^+) = 0,$$

which is equivalent to (17). The representation $X = X^+ \cup X^-$ is called the *Hahn decomposition* (of X).

Now suppose that $X \notin \mathfrak{A}$. In this case, X is the union of a sequence of sets X_n with finite measures $\mu(X_n)$, where the X_n can clearly be regarded as disjoint. For every n, let X_n^+ and X_n^+ be the sets figuring in the Hahn decomposition of X_n, and consider the two generalized Borel sets

$$X^+ = \bigcup_{n=1}^{\infty} X_n^+, \qquad X^- = \bigcup_{n=1}^{\infty} X_n^-. \tag{18}$$

If $E \subset X^+$ is a Borel set, then

$$E = \bigcup_{n=1}^{\infty} EX_n^+, \qquad \mu(E) = \sum_{n=1}^{\infty} \mu(EX_n^+) \geqslant 0,$$

while if $E \subset X^-$, then

$$E = \bigcup_{n=1}^{\infty} EX_n^-, \qquad \mu(E) = \sum_{n=1}^{\infty} \mu(EX_n^-) \leqslant 0.$$

Thus $X = X^+ \cup X^-$, with X^+ and X^- given by (18), is the required Hahn decomposition, and the proof is complete.

THEOREM 11. *Let $\mathfrak{A}$, μ, X^+ and X^- be the same as in Theorem 10. Then the positive, negative and total variations of μ are given by*

$$\lambda(E) = \mu(EX^+), \qquad \nu(E) = -\mu(EX^-) \tag{19}$$

and

$$\rho(E) = \sup \sum_{k=1}^{m} |\mu(A_k)| \tag{20}$$

for every Borel set E, where the least upper bound in (20) *is taken with respect to all finite unions $F = A_1 \cup \cdots \cup A_m$ of disjoint sets $A_k \in \mathfrak{A}$, $A_k \subset E$.*

Proof. Recalling the formulas (14), we see that $\lambda(E) = \mu(E)$, $\nu(E) = 0$ for every Borel set $E \subset X^+$, while $\lambda(E) = 0$, $\nu(E) = -\mu(E)$ for every Borel set $E \subset X^-$. But any $E \subset X$ can be written in the form

$$E = EX^+ \cup EX^-.$$

Therefore

$$\lambda(E) = \lambda(EX^+) = \mu(EX^+), \qquad \nu(E) = \nu(EX^-) = -\mu(EX^-),$$

which agrees with (19). Moreover, since

$$|\mu(A_k)| = |\lambda(A_k) - \nu(A_k)| \leqslant \lambda(A_k) + \nu(A_k) = \rho(A_k),$$

we have

$$\sum_{k=1}^{m} |\mu(A_k)| \leqslant \sum_{k=1}^{m} \rho(A_k) \leqslant \rho(E),$$

and hence

$$\sup \sum_{k=1}^{m} |\mu(A_k)| \leqslant \rho(E). \tag{21}$$

On the other hand, as just shown,

$$\rho(E) = \lambda(E) + \nu(E) = \mu(EX^+) - \mu(EX^-) = |\mu(EX^+)| + |\mu(EX^-)|,$$

and hence obviously

$$\rho(E) \leqslant \sup \sum_{k=1}^{m} |\mu(A_k)|. \tag{22}$$

Comparing (21) and (22), we obtain (20) as required.

*8.7. The General Continuous Linear Functional on the Space $C(X)$

We now extend the considerations of Sec. 5.4 to the case where the basic block **B** is replaced by a general *compact metric space* X, i.e., a space X equipped with a metric ρ such that every infinite subset of X contains a sequence converging to a point in X.[5] Let μ be a (signed) Borel measure defined on a σ-ring $\mathfrak{A}$ of subsets of X, containing X and all its open subsets, and let $f(x)$ be continuous on X. Given any real c, the set $\{x: f(x) > c\}$ is μ-measurable (being open), and hence $f(x)$ is itself μ-measurable. Moreover, $f(x)$ is μ-measurable (being μ-measurable and bounded). Therefore we can form the Lebesgue integral

$$I_\mu f = \int_X f(x)\mu(dx)$$

of any function $f(x)$ continuous on X. Let $C(X)$ be the normed linear space of all functions continuous on X, equipped with the norm

$$\|f\| = \max_{x \in X} |f(x)|$$

(cf. p. 94). Then the integral $I_\mu f$ defines a continuous linear functional on $C(X)$, since it satisfies the following two conditions:

1) If f_1, f_2 are any two functions in $C(X)$ and α_1, α_2 are any two real numbers, then
$$I_\mu(\alpha_1 f_1 + \alpha_2 f_2) = \alpha_1 I_\mu f_1 + \alpha_2 I_\mu f_2.$$

2) If $f_m \in C(X)$ is a sequence such that $\|f_m\| \to 0$ as $m \to \infty$, then $I_\mu f_m \to 0$, as follows at once from the estimate
$$|I_\mu f_m| = \left| \int_X f_m(x)\mu(dx) \right| \leqslant \|f_m\| \rho(X),$$
where ρ is the total variation of μ (see p. 161).

Next we prove the analogue of Theorem 1, p. 95:

THEOREM 12. *Given a continuous linear functional If defined on the space $C(X)$, there exists a Borel measure μ defined on a σ-ring $\mathfrak{A}$ of subsets of X, containing X and all its open subsets, such that*

$$If = \int_X f(x)\mu(dx). \tag{23}$$

Proof. First suppose I is nonnegative, so that $If \geqslant 0$ if $f(x) \geqslant 0$. Then, choosing $C(X)$ as the space of elementary functions and the

---

[5] Here convergence of a sequence x_m to a limit x_0 means convergence of the "distances" $\rho(x_m, x_0)$ to zero.

functional I as the elementary integral, we can construct a space L_I of I-summable functions. The only nontrivial part of this assertion, given the theory of Chap. 2, is to verify that I satisfies Axiom 3, p. 24. But according to Dini's lemma (p. 54), which generalizes at once to the case of a compact metric space, $f_m \searrow 0$ implies $f_m \to 0$ uniformly, i.e., $\|f_m\| \to 0$, and hence $If_m \to 0$. Now let $\mathfrak{A}$ be the class of I-measurable subsets of X. Clearly $\mathfrak{A}$ contains X and all its open subsets. In fact, if $G \subset X$ is open, we have $G = \{x: \varphi(x) > 0\}$ where $\varphi(x)$ is the distance between the point $x \in X$ and the set $X - G$, a function which is easily seen to be continuous. The measure

$$\mu(E) = I\chi_E(x), \qquad E \in \mathfrak{A}$$

is a Borel measure on $\mathfrak{A}$ (in fact, a Lebesgue measure, since $\mathfrak{A}$ is actually a σ_μ-ring). Moreover μ satisfies the relation (23), as required.

If the functional I takes values of either sign, then, according to Riesz's representation theorem (p. 44), we can represent I in the form

$$I = J - N,$$

where the linear functionals J and N are nonnegative and continuous in the sense that $f_m \searrow 0$ implies $Jh_m \to 0$, $Nh_m \to 0$. This time let $\mathfrak{A}$ be the class of K-measurable subsets of X, where $K = J + N$, and define the nonnegative measures

$$\lambda(E) = J\chi_E(x), \quad \nu(E) = N\chi_E(x), \quad E \in \mathfrak{A}$$

and the signed measure

$$\mu(E) = \lambda(E) - \nu(E).$$

Then (23) continues to hold with this choice of μ, and the proof is complete.

*8.8. The Lebesgue-Stieltjes Integral on an Infinite-Dimensional Space

In Chap. 5 we defined the Lebesgue-Stieltjes integral on a closed basic block

$$X = \{x \in R_n: a_1 \leqslant x_1 \leqslant b_1, \ldots, a_n \leqslant x_n \leqslant b_n\}$$

(denoted there by **B**) in Euclidean n-space R_n. The set X can be regarded as the set of all real functions $x_k = x(k)$ which are defined at n points $k = 1, \ldots, n$ and for each k take values in the interval $[a_k, b_k]$. If, for convenience, we make a preliminary change of variables transforming each interval $[a_k, b_k]$ into the unit interval $[0, 1]$, the set X has the particularly simple form

$$X = \{x \in R_n: 0 \leqslant x_1 \leqslant 1, \ldots, 0 \leqslant x_n \leqslant 1\}.$$

Suppose we now replace the finite set $\{1, \ldots, n\}$ of values of the index k by an arbitrary *infinite* set T of values of the index t. Correspondingly, we regard X as the set of all real functions $x(t)$ which are defined on T and for each $t \in T$ take values in the interval $[0, 1]$:

$$X = \{x(t): t \in T, 0 \leqslant x(t) \leqslant 1\}.$$

Such a set X will be called the "infinite-dimensional cube," the "T-dimensional cube," the "Cartesian product of T closed intervals" or simply the "T-cube." The aim of this section is to show how the Daniell scheme can be used to construct a theory of integration on X.

8.8.1. Cylinder sets, blocks and quasi-volumes. Extensions and projections. In the case of an n-dimensional cube, the concepts of passage to the limit and continuity of functions can be defined in terms of the distance between points. In the case of a T-dimensional cube, it is in general no longer possible to introduce a "natural distance." However, we can still introduce a "natural topology," i.e., given any $\varepsilon > 0$, any integer $n > 0$ and any n points $t_1, \ldots, t_n$, a set of the form

$$U = U_{x_0} = \{x(t) \in X: |x(t_k) - x_0(t_k)| < \varepsilon, k = 1, \ldots, n\}$$

is called a "neighborhood" of the "point" $x_0(t) \in X$. *The set X, equipped with this topology, is a topological space, in fact a compact Hausdorff space.* It is clear that X is a Hausdorff space, since it is easily verified that

a) Every point $x \in X$ has at least one neighborhood U_x;

b) If U_x is a neighborhood of x and x' is a point in U_x, then there is a neighborhood $U_{x'}$ of x' contained in U_x;

c) If U_x and U'_x are two neighborhoods of the same point $x \in X$, then there is a neighborhood U''_x of x contained in the intersection $U_x U'_x$;

d) Given any two distinct points $x, x' \in X$, there are neighborhoods U_x and $U_{x'}$ which are *disjoint*.

The compactness of X follows from *Tychonoff's theorem*, which asserts that *a Cartesian product of compact spaces is compact in the product topology.*[6]

A set $E \subset X$ is said to be a *cylinder set* if we can find an integer $n > 0$ and points $t_1, \ldots, t_n \in T$ such that

$$E = \{x(t) \in T: (x(t_1), \ldots, x(t_n)) \in E^n\}, \tag{24}$$

where the set E^n, called the *base* of the cylinder set E, is a subset of the n-dimensional cube[7]

$$X^n = \{\xi \in R_n: 0 \leqslant \xi_1 \leqslant 1, \ldots, 0 \leqslant \xi_n \leqslant 1\}.$$

---

[6] For the proof, see e.g., N. Dunford and J. T. Schwartz, *Linear Operators, Part I: General Theory*, Interscience Publishers, Inc., New York (1958), p. 32.

[7] To avoid confusion with functions $x(t) \in X$, we use a different letter $\xi = (\xi_1, \ldots, \xi_n)$ for the variable point in X^n.

In particular, the complement of the cylinder set (24) relative to the whole space X is itself a cylinder set:

$$X - E = \{x(t) \in X: (x(t_1), \ldots, x(t_n)) \in X^n - E^n\}.$$

Cylinder sets of the special form

$$B = \{x(t) \in X: \alpha_1 < x(t_1) \leqslant \beta_1, \ldots, \alpha_n < x(t_n) \leqslant \beta_n\} \tag{25}$$

[where the inequality $\alpha_j < x(t_j)$ is replaced by the equality $0 \leqslant x(t_j)$ if $\alpha_j = 0$] are called *blocks*. Since every block is specified by only a finite number of conditions, there is an integer $N > 0$ such that N conditions are enough to specify every block in a given finite family of blocks. It follows that the family of all blocks $B \subset X$ is a semiring, and, by the same token, that the family of all cylinder sets $E \subset X$ is a ring.

By a *quasi-volume* defined on X we mean an additive real set function $\omega(B)$ defined on *every* block $B \subset X$. The quasi-volumes considered here will be assumed to have two further properties:

1) *Nonnegativity and boundedness*. For every $B \subset X$,

$$\omega(B) \geqslant 0,$$

and moreover $\omega(X) < \infty$.

2) *Continuity on the empty set in every n-dimensional cube* (cf. p. 100). If $B_1 \supset B_2 \supset \cdots$ is a sequence of blocks defined by the same fixed set of coordinates $t_1, \ldots, t_n \in T$, with an empty intersection

$$\bigcap_{m=1}^{\infty} B_m = \varnothing,$$

then

$$\lim_{m\to\infty} \omega(B_m) = 0.$$

It should be emphasized that $\omega(B)$ depends only on the block B itself, and not on the particular form (25) in which B is written. Thus the quasi-volume ω must satisfy the following "consistency condition":

$$\begin{aligned} &\omega\{x: \alpha_1 < x(t_1) \leqslant \beta_1, \ldots, \alpha_n < x(t_n) \leqslant \beta_n\} \\ &\quad = \omega\{x: \alpha_1 < x(t_1) \leqslant \beta_1, \ldots, \alpha_n < x(t_n) \leqslant \beta_n, 0 \leqslant x(t_{n+1}) \leqslant 1\}. \end{aligned} \tag{26}$$

Given a quasi-volume ω defined on all blocks (25), where n and $t_1, \ldots, t_n$ are arbitrary, we can use the formula

$$\begin{aligned} &\omega^{t_1,\ldots,t_n}\{\xi \in X^n: \alpha_1 < \xi_1 \leqslant \beta_1, \ldots, \alpha_n < \xi_n \leqslant \beta_n\} \\ &\qquad = \omega\{x \in X: \alpha_1 < x(t_1) \leqslant \beta_1, \ldots, \alpha_n < x(t_n) \leqslant \beta_n\} \end{aligned} \tag{27}$$

to define a quasi-volume $\omega^{t_1,\ldots,t_n}$ on every n-dimensional cube X^n. In particular,

$$\omega^{t_1,\ldots,t_n}(X^n) = \omega(X), \tag{28}$$

and the consistency condition now takes the form

$$\omega^{t_1,\ldots,t_n}\{\xi \in X^n: \alpha_1 < \xi_1 \leqslant \beta_1, \ldots, \alpha_n < \xi_n \leqslant \beta_n\}$$
$$= \omega^{t_1,\ldots,t_n,t_{n+1}}\{\xi \in X^{n+1}: \alpha_1 < \xi_1 \leqslant \beta_1, \ldots, \alpha_n < \xi_n \leqslant \beta_n, 0 \leqslant \xi_{n+1} \leqslant 1\}.$$

It follows from Condition 2 that every quasi-volume $\omega^{t_1,\ldots,t_n}$ is continuous in the sense of Sec. 5.6.

Conversely, let

$$\omega^{t_1,\ldots,t_n} \qquad (n, t_1, \ldots, t_n \text{ arbitrary})$$

be a family of continuous nonnegative quasi-volumes, which are defined on every n-dimensional cube X^n and satisfy the consistency condition (26). Then we can define a quasi-volume ω on the set X by the simple expedient of reading (27) from right to left.

A key concept in the present theory is that of a function $f(x)$, $x \in X$ which effectively depends on only a finite number of "coordinates" $x(t_1), \ldots, x(t_n)$, i.e., whose value remains the same if any of the coordinates $x(t)$, $t \neq t_1, \ldots, t_n$ is changed. With every such function $f(x)$ we can associate a function

$$f(\xi) = f(\xi_1, \ldots, \xi_n) = f[x(t_1), \ldots, x(t_n)] \tag{29}$$

defined on the n-dimensional cube X^n. This correspondence between $f(x)$ and $f(\xi)$ is one-to-one, in the sense that given any function $f(\xi)$ defined on X^n and any n points $t_1, \ldots, t_n$ (n arbitrary), we can define a function $f(x)$ depending only on the coordinates $x(t_1), \ldots, x(t_n)$, by merely reading (29) from right to left. The function $f(x)$ will be called the *extension* of $f(\xi)$ from X^n onto X, while the function $f(\xi)$ will be called the *projection* of $f(x)$ from X onto X^n.

LEMMA 1. *If $f(\xi)$ is continuous on X^n, then $f(x)$ is continuous on X, and conversely.*

Proof. If $f(\xi)$ is continuous at $\xi_0 = (\xi_1^0, \ldots, \xi_n^0) \in X^n$, then, given any $\varepsilon > 0$, there is a δ such that

$$|\xi_k - \xi_k^0| < \delta \qquad (k = 1, \ldots, n)$$

implies

$$|f(\xi) - f(\xi_0)| = |f(x) - f(x_0)| < \varepsilon,$$

where x_0 is any point in X such that $x(t_k) = \xi_k$ ($k = 1, \ldots, n$). In other words, $|f(x) - f(x_0)| < \varepsilon$ provided that x belongs to the neighborhood

$$\{x(t) \in X: |x(t_k) - x_0(t_k)| < \delta, k = 1, \ldots, n\}$$

(recall the "natural topology" of p. 168). The converse follows by detailed reversal of steps. The fact that

$$X = \{x(t) \in X: x(t_1) = \xi_1, \ldots, x(t_n) = \xi_n, \xi \in X^n\},$$
$$X^n = \{\xi \in X^n: \xi_1 = x(t_1), \ldots, \xi_n = x(t_n), x(t) \in X\}$$

is used to go from point continuity to continuity on the sets X and X^n.

8.8.2. Construction of the space $L_\omega(X)$. Kolmogorov's theorem. We are now ready to use the Daniell scheme to construct a theory of Lebesgue-Stieltjes integration on X, with respect to a given quasi-volume ω. As our space of elementary functions we take the set H of all continuous functions on X which depend only on a finite number of coordinates. This choice of H clearly satisfies Axioms a and b, p. 23. In fact, Axiom b is obvious, while to verify Axiom a, we need only note that if $f(x)$ and $g(x)$ are two functions in H, the first depending on coordinates $x(t_1), \ldots, x(t_n)$ and the second on coordinates $x(t_1'), \ldots, x(t_r')$, then the linear combination $\alpha f(x) + \beta g(x)$, where α and β are arbitrary real numbers, also depends on only finitely many coordinates, more exactly on the coordinates $x(t_1''), \ldots, x(t_s'')$ where[8]

$$\{t_1'', \ldots, t_s''\} = \{t_1, \ldots, t_n\} \cup \{t_1', \ldots, t_r'\}.$$

As in Sec. 5.1, a natural choice of the elementary integral on H is the *finite-dimensional* Riemann-Stieltjes integral[9]

$$I_\omega f = \int_{X^n} f(\xi_1, \ldots, \xi_n)\omega^{t_1, \ldots, t_n}(d\xi), \tag{30}$$

which can also be written in the form

$$I_\omega f = \int_{X^{n+1}} f(\xi_1, \ldots, \xi_n)\omega^{t_1, \ldots, t_n, t_{n+1}}(d\xi\, d\xi_{n+1}), \tag{31}$$

because of the consistency condition (26). However, before we can invoke the general Daniell scheme of Chap. 2, we must first prove

THEOREM 13. *The integral* (30) *has all the properties of an elementary integral.*

Proof. Axiom 2, p. 24 is obvious, i.e., if $f(x) \geqslant 0$, then $I_\omega f \geqslant 0$, but it takes a little more thought to verify Axioms 1 and 3. Let $f(x)$, $g(x) \in H$ be the same as in the verification of Axiom a above, and let α, β be arbitrary real numbers. Then, because of the obvious generalization of (31),[10]

$$\begin{aligned} I_\omega(\alpha f + \beta g) &= \int_{X^s}[\alpha f(\xi'') + \beta g(\xi'')]\omega^{t_1'', \ldots, t_s''}(d\xi'') \\ &= \alpha\int_{X^s} f(\xi'')\omega^{t_1'', \ldots, t_s''}(d\xi'') + \beta\int_{X^s} g(\xi'')\omega^{t_1'', \ldots, t_s''}(d\xi'') \\ &= \alpha\int_{X^n} f(\xi)\omega^{t_1, \ldots, t_n}(d\xi) + \beta\int_{X^r} g(\xi')\omega^{t_1', \ldots, t_r'}(d\xi') \\ &= \alpha I_\omega f + \beta I_\omega g, \end{aligned}$$

[8] Note that in general $s \neq n + r$, since the sets $\{t_1, \ldots, t_n\}$ and $\{t_1', \ldots, t_r'\}$ may intersect.
[9] The existence of (30) follows from Lemma 1 and Theorem 1, p. 66.
[10] Here, of course, $\xi = (\xi_1, \ldots, \xi_n)$, $\xi' = (\xi_1', \ldots, \xi_r')$ and $\xi'' = (\xi_1'', \ldots, \xi_s'')$.

which proves Axiom 1. As for Axiom 3, we start from the estimate

$$|I_\omega f_m| \leqslant \omega^{t_1,\ldots,t_n}(X^n) \max_{\xi \in X^n} |f_m(\xi)| = \omega(X) \max_{x \in X} |f_m(x)|$$

implied by (30) and the corresponding estimate for finite-dimensional Riemann-Stieltjes integrals [formula (12), p. 68]. But according to Dini's lemma (p. 54), which generalizes at once to the case of a compact topological space, $f_m \searrow 0$ implies $f_m \to 0$ uniformly, i.e.,

$$\max_{x \in X} |f(x)| \to 0,$$

and hence $If_m \to 0$, as required.

Thus all the prerequisites for constructing a theory of the integral, based on continuous functions depending only on finitely many coordinates $x(t_1), \ldots, x(t_n)$ as elementary functions, with the integral (30) as elementary integral, are satisfied. This leads to a space $L_\omega(X)$ of ω-summable functions on X, and a corresponding σ_ω-ring of ω-summable sets. In fact, we have arrived at the following celebrated result, proved by Kolmogorov in 1933:

THEOREM 14 (*Kolmogorov's theorem*). *Let X be the infinite-dimensional cube, and let ω be a quasi-volume defined on all blocks $B \subset X$, such that every quasi-volume $\omega^{t_1,\ldots,t_n}$ is continuous on the corresponding finite-dimensional cube X^n. Then ω can be extended to a Lebesgue measure on a σ_ω-ring of subsets of X.*

Proof. We need only show that the measure $\omega(E) = I_\omega \chi_E$ defined on the σ_ω-ring of ω-measurable subsets of X reduces to the original quasi-volume ω, i.e., that

$$\omega(B) = I_\omega \chi_B(x)$$

for every block $B \subset X$. Suppose B is specified by n conditions involving the coordinates $x(t_1), \ldots, x(t_n)$. Let $I_{\omega^{t_1,\ldots,t_n}}$ be the Lebesgue-Stieltjes integral for the space of $\omega^{t_1,\ldots,t_n}$-summable functions, constructed via the Daniell scheme, starting from the space of continuous functions on X^n equipped with the Riemann-Stieltjes integral with respect to $\omega^{t_1,\ldots,t_n}$. Moreover, let $f_m(x)$ be a sequence of continuous functions on X^n converging everywhere to the characteristic function $\chi_{B^n}(x)$ of the base B^n of the block B. Then, invoking the continuity of $\omega^{t_1,\ldots,t_n}$ on X^n (cf. Sec. 5.6), we have

$$\omega(B) = \omega^{t_1,\ldots,t_n}(B^n) = I_{\omega^{t_1,\ldots,t_n}} \chi_{B^n}(\xi) = \lim_{m \to \infty} I_{\omega^{t_1,\ldots,t_n}} f_m(\xi)$$

$$= \lim_{m \to \infty} I_\omega f_m(x) = I_\omega \chi_B(x),$$

and the proof is complete.

Remark. Theorem 14 has an obvious generalization to the case of a *signed* quasi-volume ω, provided that the total variations of $\omega^{t_1,\ldots,t_n}$ are all bounded by a fixed constant, i.e., that

$$V_{\omega^{t_1,\ldots,t_n}}(X) \leqslant M$$

for arbitrary $n, t_1, \ldots, t_n$.

THEOREM 15. *The integral of a function $f(x) \in L_\omega(X)$ can be expressed as the limit of a sequence of finite-dimensional integrals, i.e.,*

$$\int_X f(x)\omega(dx) = \lim_{m\to\infty} \int_{X^{n_m}} f_m(\xi_1, \ldots, \xi_{n_m})\omega^{t_1,\ldots,t_{n_m}}(d\xi). \tag{32}$$

Proof. The set of elementary functions H, i.e., the set of all continuous functions on X depending on only finitely many coordinates, is dense in $L_\omega(X)$ [see p. 40], thereby implying (32).

COROLLARY. *The set of polynomials with (all possible) arguments $t_1, \ldots, t_n$ is dense in $L_\omega(X)$.*

Proof. Any function in H can be uniformly approximated by polynomials (Weierstrass' theorem). But H is dense in $L_\omega(X)$, as just noted.

8.8.3. Structure of ω-measurable sets and functions. As shown on p. 144, the semiring of all blocks $B^n \subset X^n$ is completely sufficient. This result is easily generalized to the infinite-dimensional case:

THEOREM 16. *The semiring $\mathfrak{A}$ of all blocks $B \subset X$ is completely sufficient.*

Proof. Given any elementary function $f(x)$, let $f(\xi)$ be its projection from X onto X^n. Then there is a sequence of step functions

$$h_m(\xi) = \sum_{k=1}^{r_m} \alpha_{km}\chi_{B^n_{km}}(\xi)$$

converging to $f(\xi)$, where the B^n_{km} are blocks in X^n. Therefore the sequence

$$h_m(x) = \sum_{k=1}^{r_m} \alpha_{km}\chi_{B_{km}}(x), \qquad B_{km} \in \mathfrak{A},$$

where B_{km} is the block with base B^n_{km}, converges to $f(x)$. This shows that $\mathfrak{A}$ is sufficient (cf. Sec. 7.3).

Next let $Z \subset X$ be a set of ω-measure zero. Then, given any $\varepsilon > 0$, there exists a nondecreasing sequence of nonnegative elementary functions $f_m(x)$ such that $I_\omega f_m < \varepsilon$ and $\sup_m f_m(x) \geqslant 1$ on X. Consider the sets

$$E_m = \{x : f_m(x) > \tfrac{1}{2}\} \qquad (m = 1, 2, \ldots).$$

Clearly $E_1 \subset E_2 \subset \cdots$, and every E_m is a cylinder set. The base E_m^n of E_m is open in X^n, being the preimage of an open set under a continuous mapping. Therefore E_m^n can be represented as a countable union of blocks in X^n, i.e.,

$$E_m^n = \bigcup_{k=1}^{\infty} B_{km}^n$$

(cf. p. 145). By the same token, E_m is open in X and can be represented as a countable union of blocks in X. In fact,

$$E_m = \bigcup_{k=1}^{\infty} B_{km},$$

where B_{km} is the block with base B_{km}^n. But obviously

$$Z \subset \bigcup_{m=1}^{\infty} E_m = \bigcup_{m=1}^{\infty} \bigcup_{k=1}^{\infty} B_{km},$$

and moreover, according to the corollary on p. 118,

$$\omega\left(\bigcup_{m=1}^{\infty} \bigcup_{k=1}^{\infty} B_{km}\right) = \lim_{m\to\infty} \omega\left(\bigcup_{k=1}^{\infty} B_{km}\right) = \lim_{m\to\infty} \omega(E_m) \leqslant 2\varepsilon,$$

since $I_\omega f_m < \varepsilon$ implies $\omega(E_m) < 2\varepsilon$. Therefore Z is covered by countably many blocks B_{km}, whose union has arbitrarily small ω-measure. In other words, $\mathfrak{A}$ is completely sufficient (cf. Sec. 7.4), and the theorem is proved.

LEMMA 2. *Given any block*

$$B = \{x(t) \in X\colon \alpha_1 < x(t_1) \leqslant \beta_1, \ldots, \alpha_n < x(t_n) \leqslant \beta_n\}$$

and any $\varepsilon > 0$, *there is a block* B' *such that every point of* B *is an interior point of* B' *and*

$$\omega(B') < \omega(B) + \varepsilon. \tag{33}$$

Proof. If B' is of the form

$$B' = \{x(t) \in X\colon \alpha_1 < x(t_1) \leqslant \beta_1', \ldots, \alpha_n < x(t_n) \leqslant \beta_n'\},$$

where $\beta_k' = \beta_k$ if $\beta_k = 1$ and $\beta_k' > \beta_k$ if $\beta_k < 1$ $(k = 1, \ldots, n)$, then every point of B is an interior point of B'. [Note that the intersection of B' with any sheet $x(t_k) = 0$ or $x(t_k) = 1$ consists entirely of interior points of B' (relative to the cube X).] Moreover,

$$B \subset B' \subset B \cup \{x(t) \in X\colon \beta_1 < x(t_1) \leqslant \beta_1'\} \cup \cdots \cup \{x(t) \in X\colon \beta_n < x(t_n) \leqslant \beta_n'\},$$

and hence

$$\omega(B) \leqslant \omega(B') \leqslant \omega(B) + \omega\{x(t) \in X\colon \beta_1 < x(t_1) \leqslant \beta_1'\} + \cdots + \omega\{x(t) \in X\colon \beta_n < x(t_n) \leqslant \beta_n'\}. \tag{34}$$

According to Condition 2, p. 169,

$$\lim_{\beta_k' \to \beta_k^+} \omega\{x(t) \in X : \beta_k < x(t_k) \leqslant \beta_k'\} = 0 \qquad (k = 1, \ldots, n),$$

i.e., given any $\varepsilon > 0$,

$$\omega\{x(t) \in X : \beta_k < x(t_k) \leqslant \beta_k'\} < \frac{\varepsilon}{n} \qquad (k = 1, \ldots, n),$$

provided that β_k' is sufficiently close to β_k. But with such a choice of the β_k', (34) implies (33), as required.

Our next result concerns the relation between ω-measurability and $\omega^{t_1, \ldots, t_n}$-measurability:

THEOREM 17. *The cylinder set $E \subset X$ is ω-measurable*[11] *if and only if its base $E^n \subset X^n$ is $\omega^{t_1, \ldots, t_n}$-measurable, and then*

$$\omega^{t_1, \ldots, t_n}(E^n) = \omega(E).$$

Proof. First we note that if $Z^n \subset X^n$ has $\omega^{t_1, \ldots, t_n}$-measure zero, then the cylinder set $Z \subset X$ with base Z^n has ω-measure zero. In fact, given any $\varepsilon > 0$, there exists a nondecreasing sequence of nonnegative functions $f_m^{(\varepsilon)}(\xi)$ continuous on X^n such that

$$I_{\omega^{t_1, \ldots, t_n}} f_m^{(\varepsilon)}(\xi) < \varepsilon, \qquad \sup_m f_m(\xi) \geqslant 1 \quad \text{on} \quad Z^n.$$

Let $f_m^{(\varepsilon)}(x)$ be the extension of $f_m^{(\varepsilon)}(\xi)$ from X^n onto X. Then $f_m^{(\varepsilon)}(x)$ is a nondecreasing sequence of nonnegative elementary functions such that

$$I_\omega f_m^{(\varepsilon)}(x) < \varepsilon, \qquad \sup_m f_m^{(\varepsilon)}(x) \geqslant 1 \quad \text{on } Z,$$

and hence Z has ω-measure zero, as asserted.

Next let $E^n \subset X^n$ be any $\omega^{t_1, \ldots, t_n}$-measurable set. Then there exists a sequence of functions $f_m(\xi)$ continuous on X^n converging on a set of full $\omega^{t_1, \ldots, t_n}$-measure to the characteristic function $\chi_{E^n}(\xi)$. Let $f_m(x)$ be the extension of $f_m(\xi)$ from X^n onto X. Then $f_m(x)$ is a sequence of elementary functions converging on a set of full ω-measure (as just shown) to the characteristic function $\chi_E(x)$, and hence E is ω-measurable. Moreover, the functions $f_m(\xi)$ and $f_m(x)$ can be regarded as bounded, and hence, by Lebesgue's theorem,

$$\omega^{t_1, \ldots, t_n}(E^n) = \lim_{m \to \infty} I_{\omega^{t_1, \ldots, t_n}} f_m(\xi) = \lim_{m \to \infty} I_\omega f_m(x) = \omega(E),$$

and half of the theorem is proved.

---

[11] And hence ω-summable, since $\omega(X) < \infty$.

The converse is more difficult. Suppose the (arbitrary) cylinder set $E \subset X$ with base $E^n \subset X^n$ is ω-measurable. To show that E^n is $\omega^{t_1, \ldots, t_n}$-measurable, it will be enough to prove that

$$\omega^{*t_1, \ldots, t_n}(E^n) \leqslant \omega(E), \tag{35}$$

where $\omega^{*t_1, \ldots, t_n}(E^n)$ denotes the outer measure of E^n (see Sec. 7.5). In fact, if (35) holds, then

$$\omega^{*t_1, \ldots, t_n}(X^n - E^n) \leqslant \omega(X - E),$$

since $X - E$ is a cylinder set with base $X^n - E^n$, and hence

$$\begin{aligned}\omega^{*t_1, \ldots, t_n}(E^n) + \omega^{*t_1, \ldots, t_n}(X^n - E^n) &\leqslant \omega(E) + \omega(X - E)\\ &= \omega(X) = \omega^{t_1, \ldots, t_n}(X^n)\end{aligned}$$

[cf. (28)], which, according to Theorem 6, p. 142, implies that E^n is $\omega^{t_1, \ldots, t_n}$-measurable.[12]

Thus we now concentrate our attention on proving the inequality (35). Given any $\varepsilon > 0$, Theorem 16 guarantees the existence of a countable set of blocks B_m such that

$$E \subset \bigcup_{m=1}^{\infty} B_m, \qquad \sum_{m=1}^{\infty} \omega(B_m) < \omega(E) + \varepsilon.$$

In general, there are points of E which are not interior points of $B_1 \cup B_2 \cup \cdots$, a fact which would prevent subsequent use of the finite subcovering lemma (p. 13). To avoid this difficulty, we use Lemma 2 to replace every block B_m by a block B'_m such that every point of B_m is an interior point of B'_m and

$$\sum_{m=1}^{\infty} \omega(B'_m) < \omega(E) + 2\varepsilon.$$

To keep the notation simple, we imagine that this replacement has already been made, and henceforth omit primes. Every block B_m is specified by a finite number of conditions involving the coordinates $x(t_k)$. These conditions can always be regarded as including conditions on the coordinates $x(t_1), \ldots, x(t_n)$ defining the cube X^n, if we allow redundant inequalities of the form $0 \leqslant x(t_k) \leqslant 1$, $k = 1, \ldots, n$. Thus every block

$$B_m = \{x(t) \in X\colon \alpha_1 < x(t_1) \leqslant \beta_1, \ldots, \alpha_n < x(t_n) \leqslant \beta_n, \ldots\}$$

---

[12] Equation (16), p. 142 can also be written in the form

$$\mu^*(E) + \mu^*(\mathscr{C}E) \leqslant \mu(X),$$

since the inequality is in any event impossible.

has a corresponding "projection"

$$B_m^n = \{\xi \in X^n : \alpha_1 < \xi_1 \leqslant \beta_1, \ldots, \alpha_n < \xi_n \leqslant \beta_n\}$$

onto the cube X^n.

Now let $\tilde{\xi} = (\tilde{\xi}_1, \ldots, \tilde{\xi}_n)$ be a fixed point in E^n, and consider the set $Q(\tilde{\xi})$ of all $x(t) \in X$ such that

$$x(t_k) = \tilde{\xi}_k \qquad (k = 1, \ldots, n).$$

Since $Q(\tilde{\xi})$ is a compact subset of X, it has the finite subcovering property. Therefore from the blocks $B_1, B_2, \ldots$ covering $Q(\tilde{\xi})$ we can select a finite set $\tilde{B}_1, \ldots, \tilde{B}_r$ covering $Q(\tilde{\xi})$, with projections $\tilde{B}_1^n, \ldots, \tilde{B}_r^n$ onto X^n. Clearly, $\tilde{\xi}$ is an interior point of the block

$$\tilde{B}^n = \bigcap_{k=1}^r \tilde{B}_k^n,$$

and also of a subblock $\underline{\tilde{B}}^n \subset \tilde{B}^n$ whose boundary sheets all have rational coordinates. Let $\underline{\tilde{B}}_1, \ldots, \underline{\tilde{B}}_r$ be the subblocks of $\tilde{B}_1, \ldots, \tilde{B}_r$ whose projections onto X^n coincide with $\underline{B}^n$, but whose coordinates other than $x(t_1), \ldots, x(t_n)$ are the same as those of $\tilde{B}_1, \ldots, \tilde{B}_r$. Then the union

$$\bigcup_{k=1}^r \underline{\tilde{B}}_k$$

is a cylinder set (in fact, a block) with base $\underline{\tilde{B}}^n$. Carrying out the same procedure for every point $\xi \in E^n$, we succeed in covering the cylinder set E by a cylinder set G, which is a union of blocks with "rational projections" onto X^n. Therefore G is a countable union of blocks, and hence an ω-measurable (Borel) set in X. Moreover

$$\omega(G) \leqslant \sum_{m=1}^{\infty} \omega(B_m) < \omega(E) + 2\varepsilon,$$

since every point of G belongs to one of the original blocks B_m. Let G^n be the (Borel) base of G. By the first half of the theorem,

$$\omega^{t_1, \ldots, t_n}(G^n) = \omega(G).$$

But then

$$\omega^{*t_1, \ldots, t_n}(E^n) \leqslant \omega^{t_1, \ldots, t_n}(G^n) = \omega(G) < \omega(E) + 2\varepsilon,$$

by the definition of outer measure ($E \subset G$). Taking the limit as $\varepsilon \to 0$, we obtain (35), and the theorem is proved.

COROLLARY 1. *A function $f(x)$, $x \in X$ depending on only a finite number of coordinates $x(t_1), \ldots, x(t_n)$ is ω-measurable if and only if its projection $f(\xi)$ from X to X^n is $\omega^{t_1, \ldots, t_n}$-measurable, and then*

$$\omega\{x(t) \in X : f(x) > c\} = \omega^{t_1, \ldots, t_n}\{\xi \in X^n : f(\xi_1, \ldots, \xi_n) > c\} \quad (36)$$

for arbitrary real c.

Proof. According to Theorem 5, p. 120, ω-measurability of $f(x)$ is equivalent to ω-measurability of every set

$$\{x(t) \in X: f(x) > c\}, \tag{37}$$

while $\omega^{t_1, \ldots, t_n}$-measurability of $f(\xi)$ is equivalent to $\omega^{t_1, \ldots, t_n}$-measurability of every set

$$\{\xi \in X^n: f(\xi) > c\}. \tag{38}$$

But, because of the nature of $f(x)$, (37) is a cylinder set with base (38), and the rest of the proof follows from Theorem 17.

COROLLARY 2. *A function $f(x)$, $x \in X$ depending on only a finite number of coordinates $x(t_1), \ldots, x(t_n)$ is ω-summable if and only if its projection $f(\xi)$ from X to X^n is $\omega^{t_1, \ldots, t_n}$-summable, and then*

$$\int_X f(x)\omega(dx) = \int_{X^n} f(\xi_1, \ldots, \xi_n)\omega^{t_1, \ldots, t_n}(d\xi).$$

Proof. We need only use (36), recalling the construction of the Lebesgue integral given in Secs. 6.6 and 8.3.

PROBLEMS

1 ("Explicit construction of a nonmeasurable set"). On the unit square $X = \{x: 0 \leqslant x_1 \leqslant 1, 0 \leqslant x_2 \leqslant 1\}$ consider the ring of sets $E = E_1 \times [0, 1]$, where $E_1 \subset [0, 1]$ is Lebesgue-measurable with measure $m(E_1)$. Then write $\mu(E) = m(E_1)$. Show that the set $[0, 1] \times [0, \frac{1}{2}]$ is μ-nonmeasurable.

Comment. Despite this problem, it is difficult to give an explicit construction of a nonmeasurable set in a general Hausdorff space all of whose open sets are measurable.

2. Let μ be a Lebesgue measure on the real line, and let $Y \subset X$ be a nonmeasurable set. Then $\mu_*(Y) = \alpha$, $\mu^*(Y) = \beta$, where $\alpha < \beta$. Construct a Lebesgue extension ν of the measure μ in which the set Y is ν-measurable with a given measure $\gamma = \nu(Y)$, $\alpha < \gamma < \beta$.

Hint. If $\mu^*(X) = \beta$, $\mu^*(Y) = 0$, the μ-measurable sets A and B are uniquely determined (to within sets of measure zero) by the set $C = AY \cup B(X - Y)$. Let $\mathfrak{A}$ (the domain of ν) consist of all such sets C, equipped with the measure

$$\nu(C) = \frac{\gamma}{\beta}\mu(A) + \left(1 - \frac{\gamma}{\beta}\right)\mu(B).$$

In the general case, there are μ-measurable sets E_1 and E_2 such that $E_1 \subset Y \subset E_2$, $\mu(E_1) = \alpha$, $\mu(E_2) = \beta$. Then apply the same construction to the difference $E = E_2 - E_1$, i.e., let $\mathfrak{A}$ include all subsets of the set E of the form

$$C = A(Y - E_1) \cup B(E_2 - Y), \qquad A \subset E, \quad B \subset E,$$

where A, B are μ-measurable and

$$\nu(C) = \frac{\gamma - \alpha}{\beta - \alpha}\,\mu(A) + \left(1 - \frac{\gamma - \alpha}{\beta - \alpha}\right)\mu(B).$$

Comment. The process of constructing Lebesgue extensions can be continued by adjunction of further nonmeasurable sets $Y_2, Y_3, \ldots$, besides $Y = Y_1$. This naturally raises the question of whether it is possible, by using transfinite induction, to extend the measure μ, as a countably additive set function, onto the family of *all* subsets of X. An extension of μ is indeed possible, but with preservation of finite additivity only; countable additivity is lost on passage to the first uncountable ordinal. In fact, there does not exist a countably additive set function defined on all subsets of the interval $[0, 1]$ equal to zero on every set containing only one point.[14]

3. Let μ be an additive set function defined on a semiring $\mathfrak{A}$ (μ need not be nonnegative). Prove that μ can always be extended to an additive set function on some ring $\mathfrak{B}$ containing $\mathfrak{A}$.

Hint. Let $\mathfrak{B}$ consist of all finite unions of disjoint sets of $\mathfrak{A}$, and then define

$$\mu(B) = \sum_m \mu(A_m) \text{ for every } B = \bigcup_m A_m \in \mathfrak{B}.$$

Verify the uniqueness and additivity of μ.

4. Suppose a countably additive nonnegative measure μ is extended from a semiring $\mathfrak{A}$ to a ring $\mathfrak{B}$, as in the preceding problem. Prove that the countable additivity of μ is preserved.

Hint. A sequence $B_1, B_2, \ldots$ of disjoint sets of $\mathfrak{B}$ can be written as

$$B_1 = A_1 \cup \cdots \cup A_{p_1}, \qquad B_2 = A_{p_1+1} \cup \cdots \cup A_{p_2}, \ldots,$$

in terms of disjoint sets of $\mathfrak{A}$.

5. Show that the three equivalent characterizations of countable additivity for a ring, given by Properties a, b and c of Sec. 8.1, are not equivalent for a semiring.

Hint. Consider the semiring consisting of all sets of rational numbers in the interval $[0, 1]$ satisfying all possible inequalities of the form $\alpha < x < \beta$, $\alpha \leqslant x < \beta$, $\alpha < x \leqslant \beta$, $\alpha \leqslant x \leqslant \beta$, where $\alpha = \beta$ is not excluded, and introduce the measure $\mu(A) = \beta - \alpha$. Property a is not satisfied (the measure is additive, but not countably additive), although Property b holds.

6. Specialize the considerations of Sec. 8.5 to the case $n = 1$.

Hint. Cf. Sec. 4.7.4 and the remark on p. 105.

---

[14] This was shown with the use of the continuum hypothesis by S. Banach and C. Kuratowski, *Sur une généralisation du problème de la mesure*, Fundamenta Mathematicae, **14,** 127 (1929).

7.[15] Let X be an uncountably infinite set, and let $\mathfrak{A}$ be the ring consisting of all finite subsets of X and their complements. For finite $E \subset X$, let $\mu(E)$ equal the number of elements in E, and otherwise let $\mu(X - E) = -\mu(E)$. Then the measure $\mu(E)$ is countably additive on $\mathfrak{A}$, but its total variation does not exist. Why is this compatible with the results of Sec. 8.4?

Hint. $\mathfrak{A}$ is a ring, but not a σ-ring.

[15] Due to O. G. Smolyanov.

Part 4

THE DERIVATIVE

9

MEASURE AND SET FUNCTIONS

9.1. Classification of Set Functions. Decomposition into Continuous and Discrete Components

Let $\mathfrak{A}$ be a σ-ring of subsets of X, equipped with a nonnegative Borel measure μ. If $X \notin \mathfrak{A}$, we assume that $X = X_1 \cup X_2 \cup \cdots$, where $X_1 \subset X_2 \subset \cdots$ and every $X_n \in \mathfrak{A}$. Besides the measure μ, we shall consider other countably additive (finite) set functions $\Phi(E)$, $E \in \mathfrak{A}$, defined on the same σ-ring, where $\Phi(E)$ is in general *signed* (i.e., of variable sign). According to Secs. 8.2 and 8.3, the Borel measure μ can be extended from the σ-ring $\mathfrak{A}$ to a Lebesgue measure (which we continue to denote by μ) on a σ_μ-ring $\bar{\mathfrak{A}} \supset \mathfrak{A}$, which can then be used to construct a space of μ-summable functions. In general, however, the function $P(E)$ cannot be extended onto the σ_μ-ring $\bar{\mathfrak{A}}$.[1] In any event, as we know from Sec. 8.4, $\Phi(E)$ can be decomposed into positive and negative variations $\Phi(E)$ and $Q(E)$, defined on the original σ-ring $\mathfrak{A}$, i.e., $\Phi(E) = P(E) - Q(E)$, $E \in \mathfrak{A}$.

With a view to classifying set functions, we now introduce some new terminology, where in each of the following definitions, $\Phi(E)$ denotes a countably additive set function, defined on the Borel sets E of some σ-ring $\mathfrak{A}$:

1) $\Phi(E)$ is said to be *concentrated* on a set E_0 if it is defined and equal to zero on every Borel set $E \subset X - E_0$. If $\Phi(E)$ is concentrated on a

[1] According to p. 161, the function $\Phi(E)$ can be extended to a signed finitely-Lebesgue measure on some σ_Φ-ring $\mathfrak{A}^*$, but the σ_μ-ring $\bar{\mathfrak{A}}$ may not be contained in $\mathfrak{A}^*$ (see Prob. 10, p. 204).

set E_0, then so are its positive variation $P(E)$, negative variation $Q(E)$ and total variation $V(E) = P(E) + Q(E)$, since if $\Phi(E)$ vanishes on every Borel set $E \subset X - E_0$, then, as in Sec. 8.4,

$$P(E) = \sup_{A \subset E} \Phi(A) = 0, \qquad Q(E) = \sup_{A \subset E} [-\Phi(A)] = 0,$$

$$V(E) = P(E) + Q(E) \qquad (A \in \mathfrak{A}).$$

2) $\Phi(E)$ is said to be *continuous* if it is defined and equal to zero on every set E of measure zero containing a single point.
3) $\Phi(E)$ is said to be *absolutely continuous* if it is defined and equal to zero on every set E of measure zero.
4) $\Phi(E)$ is said to be *singular* if it is concentrated on a set E_0 of measure zero.
5) $\Phi(E)$ is said to be *discrete* if it is concentrated on a set E_0 of measure zero containing no more than countably many points.

Next we deduce some simple consequences of these definitions and our previous considerations:

1) *Every absolutely continuous set function is continuous.*
2) *Every discrete set function is singular.*
3) *The underlying Borel measure $\mu(E)$ is absolutely continuous.*
4) *If $\Phi(E)$ is absolutely continuous, so are its positive variation $P(E)$, negative variation $Q(E)$ and total variation $V(E) = P(E) + Q(E)$.*
5) *If $\Phi(E)$ is absolutely continuous and singular, then $\Phi(E) \equiv 0$.*
6) *If $g(x)$ is a μ-summable function, then*

$$\Phi(E) = \int_E g(x)\mu(dx)$$

is an absolutely continuous set function.

There exist continuous set functions that are not absolutely continuous. For example, let X be the square $0 \leqslant x \leqslant 1$, $0 \leqslant y \leqslant 1$, equipped with ordinary two-dimensional Lebesgue measure, and let $\Phi(E)$ be the ordinary one-dimensional Lebesgue measure of the intersection of E with the interval $0 \leqslant x \leqslant 1$. Then $\Phi(E)$ is continuous and singular, but not absolutely continuous. Some less trivial examples are given in Probs. 4 and 6, pp. 203–204.

If the set function $\Phi(E)$ is discrete, then

$$\Phi(E) = \sum_{c_m \in E} g_m \qquad (E \in \mathfrak{A}), \tag{1}$$

where $c_1, \ldots, c_m, \ldots$ is a sequence of points in X of zero measure and $g_1, \ldots, g_m, \ldots$ is a corresponding sequence of real numbers such that

$$\sum_{m=1}^{\infty} |g_m| = \sum_{c_m \in X} |g_m| < \infty \tag{2}$$

if $X \in \mathfrak{A}$.[2] If $X \notin \mathfrak{A}$, in which case $X = X_1 \cup X_2 \cup \cdots$, where $X_1 \subset X_2 \subset \cdots$ and every $X_n \in \mathfrak{A}$, we replace (2) by the condition that

$$\sum_{c_m \in X_n} |g_m| < \infty \qquad (n = 1, 2, \ldots).$$

The positive, negative and total variations $P(E)$, $Q(E)$ and $V(E)$ of a discrete set function are easily found. In fact,

$$P(E) = \sum_{\substack{c_m \in E \\ g_m > 0}} g_m, \qquad Q(E) = \sum_{\substack{c_m \in E \\ g_m < 0}} |g_m|, \quad V(E) = \sum_{c_m \in E} |g_m| .$$

THEOREM 1. *Let $\mathfrak{A}$ be a σ-ring of subsets of X, equipped with a nonnegative Borel measure $\mu(E)$ and a countably additive set function $\Phi(E)$. Then $\Phi(E)$ can be represented in the form*

$$\Phi(E) = C(E) + D(E), \tag{3}$$

where $C(E)$ is continuous and $D(E)$ is discrete.

Proof. Because of Theorem 8, p. 160, there is no loss of generality in assuming that $\Phi(E)$ is nonnegative. Since, given any $\varepsilon > 0$, no Borel set contains more than a finite number of points such that

$$\mu(\{x\}) = 0, \qquad \Phi(\{x\}) > \varepsilon,$$

the whole set X can contain no more than countably many such points. Choosing

$$\varepsilon = 1, \frac{1}{2}, \ldots, \frac{1}{m}, \ldots,$$

we see that the set of all x such that

$$\mu(\{x\}) = 0, \qquad \Phi(\{x\}) > 0$$

contains no more than countably many points. Let these points be $c_1, \ldots, c_m, \ldots$ Then, given any $E \in \mathfrak{A}$, the set function

$$D(E) = \sum_{c_m \in E} \Phi(c_m) \leqslant \Phi(E) < \infty$$

is obviously discrete and finite. By construction, the function

$$C(E) = \Phi(E) - D(E)$$

takes the value zero on any set $\{x\}$ of measure zero, and hence is continuous, thereby implying the representation (3).

---

[2] As on pp. 91, 146, (1) means the sum of all the numbers g_m such that the corresponding points c_m belong to the set E. Note that $\Phi(\{c_m\}) = g_m$.

According to the Hahn decomposition (Theorem 10, p. 163), if $\Phi(E)$ is a countably additive set function defined on a σ-ring $\mathfrak{A}$ of subsets of X, then X is the union of two generalized Borel sets X^+ and X^- such that

$$\Phi(E) \geqslant 0 \qquad \text{if} \quad E \subset X^+,$$
$$\Phi(E) \leqslant 0 \qquad \text{if} \quad E \subset X^-$$

for every $E \in \mathfrak{A}$. We now prove a related result:

THEOREM 2. *Let $\mathfrak{A}$ be a σ-ring of subsets of X, equipped with a nonnegative Borel measure $\mu(E)$ and a nonnegative countably additive set function $\Phi(E)$. Then X is the union $Z_0 \cup E_1 \cup E_2 \cup \cdots$ of a sequence of disjoint generalized Borel sets $Z_0, E_1, E_2, \ldots$ such that*

$$a(n-1)\mu(E) \leqslant \Phi(E) \leqslant an\mu(E) \tag{4}$$

if $E \subset E_n$, $E \in \mathfrak{A}$, while

$$\mu(Z) = 0$$

if $Z \subset Z_0$, $Z \in \mathfrak{A}$.

Proof. Since $\Phi(E)$ and $\mu(E)$ are countably additive, so is the set function

$$\Phi_n(E) = \Phi(E) - an\mu(E) \qquad (n = 1, 2, \ldots).$$

Let $X = X_n^+ \cup X_n^-$ be the Hahn decomposition of Φ_n. Then $\Phi(E) \geqslant an\mu(E)$ if $E \subset X_n^+$ and $\Phi(E) \leqslant 0$ if $E \subset X_n^-$. Thus (4) holds on any subset of the set $G_n = X_{n-1}^+ X_n^-$. Clearly

$$G_1 = X_0^+ X_1^- = X X_1^- = X_1^-,$$
$$G_2 = X_1^+ X_2^- = (X - X_1^-) X_2^- = X_2^- - X_1^- X_2^-, \ldots,$$
$$G_n = X_{n-1}^+ X_n^- = (X - X_{n-1}^-) X_n^- = X_n^- - X_{n-1}^- X_n^-, \ldots,$$

and moreover, the complement of $G = G_1 \cup G_2 \cup \cdots = X_1^- \cup X_2^- \cup \cdots$ (relative to X) is

$$Z_0 = X - \bigcup_n X_n^- = \bigcap_n (X - X_n^-) = \bigcap_n X_n^+.$$

If $Z \subset Z_0$, $Z \in \mathfrak{A}$, then $Z \subset X_n^+$ for every n, and hence

$$\Phi(Z) \geqslant an\mu(Z) \qquad (n = 1, 2, \ldots),$$

which implies $\mu(Z) = 0$. The sets G_n will in general intersect. However, the sets

$$E_1 = X_1^- = G_1, \qquad E_2 = X_2^- - X_1^- X_2^- = G_2, \ldots,$$
$$E_n = X_n^- - X_1^- X_n^- - \cdots - X_{n-1}^- X_n^- \subset G_n, \ldots$$

are disjoint, and (4) holds on E_n. Moreover, the E_n have the same union G as the G_n, and hence lead to the same set Z_0. This completes the proof.

9.2. Decomposition of a Continuous Set Function into Absolutely Continuous and Singular Components. The Radon-Nikodým Theorem

Next we show how to carry the decomposition (3) even further:

THEOREM 3. *Let $\mathfrak{A}$ be a σ-ring of subsets of X, equipped with a nonnegative Borel measure $\mu(E)$ and a countably additive set function $\Phi(E)$. Then $\Phi(E)$ can be represented in the form*

$$\Phi(E) = A(E) + S(E) + D(E), \tag{4}$$

where $A(E)$ is absolutely continuous, $S(E)$ is continuous and singular, and $D(E)$ is discrete. Moreover, $A(E)$ itself has the representation

$$A(E) = \int_E f(x)\mu(dx), \tag{5}$$

where the function $f(x)$ is μ-summable on every Borel set E. The representation (4) *is unique, and so is $f(x)$, to within a set of measure zero.*

Proof.[3] Again there is no loss of generality in assuming that $\Phi(E)$ is nonnegative. Moreover, because of Theorem 1, we can assume that $\Phi(E)$ is continuous, provided a discrete component $D(E)$ is added to the final answer. According to Theorem 2, for every $m = 1, 2, \ldots$ there is a decomposition

$$X = Z^{(m)} \cup E_1^{(m)} \cup E_2^{(m)} \cup \cdots$$

of the set X into disjoint generalized Borel sets $Z^{(m)}, E_1^{(m)}, E_2^{(m)}, \ldots$ (recall footnote 4, p. 163), such that

$$\frac{n-1}{2^m}\mu(E) \leqslant \Phi(E) \leqslant \frac{n}{2^m}\mu(E), \qquad \mu(Z) = 0,$$

where E is an arbitrary Borel set contained in $E_n^{(m)}$ and Z is an arbitrary Borel set contained in $Z^{(m)}$. Moreover, if

$$Z_0 = \bigcup_{m=1}^{\infty} Z^{(m)},$$

then any Borel set Z contained in Z_0 also has measure zero. Consider the function

$$f_m(x) = \frac{n-1}{2^m} \qquad \text{for} \quad x \in E_n^{(m)}, \quad m, n = 1, 2, \ldots, \tag{6}$$

[3] Following S. Saks, *Theory of the Integral* (translated by L. C. Young, with two notes by S. Banach), second revised edition, Dover Publications, Inc., New York (1964), Chap. 1, Sec. 14.

defined everywhere on X except on the set $Z^{(m)}$. Given any Borel set E, we have

$$E = EZ^{(m)} \cup EE_1^{(m)} \cup EE_2^{(m)} \cup \cdots,$$

and hence

$$\Phi(E) = \Phi(EZ^{(m)}) + \sum_{n=1}^{\infty} \Phi(EE_n^{(m)}) \geqslant \sum_{n=1}^{\infty} \frac{n-1}{2^m} \mu(EE_n^{(m)}) = \int_E f_m(x)\mu(dx), \tag{7}$$

which, in particular, implies the μ-summability of $f_m(x)$ on E. We also have

$$\begin{aligned} \Phi(E) &\leqslant \Phi(EZ_0) + \sum_{n=1}^{\infty} \frac{n}{2^m} \mu(EE_n^{(m)}) \\ &= \Phi(EZ_0) + \int_E f_m(x)\mu(dx) + \frac{1}{2^m}\mu(E). \end{aligned} \tag{8}$$

We now estimate the difference between the functions $f_m(x)$ and $f_{m+1}(x)$. Since $E_n^{(m)}E_k^{(m+1)}$ is contained in both $E_n^{(m)}$ and $E_k^{(m+1)}$,

$$\frac{n-1}{2^m}\mu(E) \leqslant \Phi(E) \leqslant \frac{n}{2^m}\mu(E),$$

$$\frac{k-1}{2^{m+1}}\mu(E) \leqslant \Phi(E) \leqslant \frac{k}{2^{m+1}}\mu(E)$$

for every Borel set $E \subset E_n^{(m)}E_k^{(m+1)}$, and hence

$$\frac{k-1}{2^{m+1}}\mu(E) \leqslant \frac{n}{2^m}\mu(E), \qquad \frac{n-1}{2^m}\mu(E) \leqslant \frac{k}{2^{m+1}}\mu(E).$$

Therefore, if $\mu(E) > 0$, we have $k - 1 \leqslant 2n$ and $2(n-1) \leqslant k$, or equivalently, $2n - 2 \leqslant k \leqslant 2n + 1$. Clearly, $\mu(E) = 0$ for other values of k. It follows that

$$E_n^{(m)} \subset E_{2n-2}^{(m+1)} \cup \cdots \cup E_{2n+1}^{(m+1)} \cup Q_n^{(m)}, \qquad \mu(Q_n^{(m)}) = 0,$$

and hence, at every point of $E_n^{(m)}$, except possibly on a set of measure zero, $f_{m+1}(x)$ takes values from

$$\frac{2n-3}{2^{m+1}} = \frac{n-1}{2^m} - \frac{1}{2^{m+1}} \quad \text{to} \quad \frac{2n}{2^{m+1}} = \frac{n-1}{2^m} + \frac{1}{2^m},$$

because of (6). Therefore the inequality

$$|f_m(x) - f_{m+1}(x)| \leqslant \frac{1}{2^m}$$

holds everywhere on $E_n^{(m)}$, except possibly on a set of measure zero. But then the functions $f_m(x)$ form a sequence converging uniformly (almost everywhere on X) to a limit function

$$f(x) = \lim_{m\to\infty} f_m(x).$$

Since, as shown above, $f_m(x)$ is μ-summable on every Borel set E, the same is true of $f(x)$, by Fatou's lemma (p. 37). Taking the limit as $m \to \infty$ in the inequalities (7) and (8), we obtain

$$\int_E f(x)\mu(dx) \leqslant \Phi(E) \leqslant \Phi(EZ_0) + \int_E f(x)\mu(dx). \tag{9}$$

Now let

$$\Phi(E) = \int_E f(x)\mu(dx) + S(E) = A(E) + S(E),$$

where the set functions $S(E)$ and $A(E)$ are obviously countably additive. Because of (9), $S(E)$ is nonnegative and satisfies the inequality $S(E) \leqslant \Phi(EZ_0)$. Therefore $S(E)$ is concentrated on the set Z_0 of measure zero, and hence is singular. As for $A(E)$, it is absolutely continuous, being the integral of a function $f(x)$ summable on every Borel set E. This proves the representation (4) and (5).

We must still prove the uniqueness of the decomposition (4) and of the function $f(x)$ figuring in (5). Suppose there are two decompositions of $\Phi(E)$ into absolutely continuous and singular components:

$$\Phi(E) = A_1(E) + S_1(E), \qquad \Phi(E) = A_2(E) + S_2(E).$$

Then

$$A_1(E) - A_2(E) = S_2(E) - S_1(E),$$

where the left-hand side is absolutely continuous, while the right-hand side is singular. But a function which is both absolutely continuous and singular must vanish, and hence $A_1 = A_2$, $S_1 = S_2$, as required. Finally, suppose there are two functions $f_1(x)$ and $f_2(x)$ such that

$$\int_E f_1(x)\mu(dx) = \int_E f_2(x)\mu(dx)$$

for every Borel set E, and let $f(x) = f_1(x) - f_2(x)$. Then

$$\int_E f(x)\mu(dx) = 0,$$

and choosing E to be the set where $f(x) > 0$, we find that $f^+(x) = 0$ almost everywhere. Similarly, $f^-(x) = 0$ almost everywhere, and hence the same is true of $f(x) = f^+(x) - f^-(x)$. In other words, the function $f(x)$ figuring in (5) is unique to within a set of measure zero, and the proof is complete.

COROLLARY (*Radon-Nikodým theorem*). *Let $\mathfrak{A}$ be a σ-ring of subsets of X equipped with a nonnegative Borel measure $\mu(E)$ and an absolutely continuous countably additive set function $\Phi(E)$. Then*

$$\Phi(E) = \int_E f(x)\mu(dx),$$

where the function $f(x)$ is μ-summable on every Borel set E and is unique to within a set of measure zero.

*9.3. Some Consequences of the Radon-Nikodým Theorem

We now draw some important conclusions from the Radon-Nikodým theorem, allowing us to continue the study of linear functionals begun in Secs. 5.5 and 8.7.

9.3.1. The general continuous linear functional on the space $L(X)$. Given a set X, let $L(X)$ be the space of all functions $f(x)$ summable on X, equipped with the norm

$$\|f\| = I(|f|) = \int_X f(x)\,dx.$$

Let $\varphi(x)$ be measurable on X, and suppose $\varphi(x)$ is also *essentially bounded* on X, i.e., suppose there is a (finite) number C such that $|\varphi(x)| \leqslant C$ on a set of full measure (relative to I). Then the integral

$$I_\varphi f = I(f\varphi) = \int_X f(x)\varphi(x)\,dx$$

exists and defines a continuous linear functional on $L(X)$, since it satisfies the following two conditions:

1) If f_1, f_2 are any two functions in $L(X)$ and α_1, α_2 are any two real numbers, then

$$I_\varphi(\alpha_1 f_1 + \alpha_2 f_2) = \alpha_1 I_\varphi f_1 + \alpha_2 I_\varphi f_2.$$

2) If $f_m \in L(X)$ is a sequence such that $\|f_m\| \to 0$ as $m \to \infty$, then $I_\varphi f_m \to 0$, as follows at once from the estimate

$$|I_\varphi f_m| = \left|\int_X f_m(x)\varphi(x)\,dx\right| \leqslant I(|f_m|) \operatorname*{ess\,sup}_{x\in X} |\varphi(x)|,$$

where

$$\operatorname*{ess\,sup}_{x\in X} |\varphi(x)|$$

denotes the greatest lower bound of all numbers C such that $|\varphi(x)| \leqslant C$ on a set of full measure.

Next we prove the analogue of Theorem 1, p. 95 and Theorem 12, p. 166:

THEOREM 4. *Given a continuous linear functional Jf defined on the space $L(X)$, there exists a measurable essentially bounded function $\varphi(x)$ such that*

$$Jf = \int_X f(x)\varphi(x)\,dx, \tag{10}$$

and moreover

$$\|J\| = \operatorname*{ess\,sup}_{x\in X} |\varphi(x)|, \tag{11}$$

where

$$\|J\| = \sup_{\|f\| \leqslant 1} |Jf|$$

is the norm of the functional.[4]

Proof. For every summable set E, with characteristic function χ_E, we write

$$\Phi(E) = J\chi_E,$$

thereby defining a countably additive set function $\Phi(E)$. Since

$$|\Phi(E)| = |J\chi_E| \leqslant \|J\| \, \|\chi_E\| = \|J\| \, I\chi_E = \|J\| \, \mu(E), \tag{12}$$

$\Phi(E)$ is absolutely continuous. Therefore, according to the Radon-Nikodým theorem,

$$\Phi(E) = \int_E \varphi(x) \, dx = I(\chi_E \varphi),$$

where $\varphi(x)$ is summable on every summable set E. Moreover, $\varphi(x)$ satisfies the inequality

$$\operatorname*{ess\,sup}_{x \in X} |\varphi(x)| \leqslant \|J\| .$$

In fact, suppose $|\varphi(x)| > \|J\|$ on a set E such that $0 < \mu(E) < \infty$, and let $E = E_+ \cup E_-$, where $E_+ = \{x: \varphi(x) > \|J\|\}$ and $E_- = \{x: \varphi(x) < -\|J\|\}$. Then at least one of the sets E_+ and E_- has positive measure. If $\mu(E_+) > 0$, we have

$$J\chi_{E_+} = \Phi(E_+) = \int_{E_+} \varphi(x) \, dx > \|J\| \, \mu(E_+),$$

while if $\mu(E_-) > 0$,

$$|J\chi_{E_-}| = |\Phi(E_-)| = \int_{E_-} |\varphi(x)| \, dx > \|J\| \, \mu(E_-).$$

On the other hand, according to (12),

$$|\Phi(E_\pm)| \leqslant \|J\| \, \mu(E_\pm).$$

---

[4] We remind the reader of the following elementary facts about the continuous linear functional J (which can take either sign, contrary to the notation of Sec. 2.11). Since J is continuous at $\|f\| = 0$, given any $\varepsilon > 0$, there is a neighborhood $\|f\| < \delta$ in which $|Jf| < \varepsilon$. But then, since J is linear,

$$\left| J\left(\frac{\delta}{2} f\right) \right| = \frac{\delta}{2} |Jf| < \frac{\delta\varepsilon}{2}$$

if $\|f\| < 1$, and hence $\|J\|$ exists. Moreover, if $\|f\| \neq 0$,

$$\left(\frac{f}{\|f\|}\right) = \frac{1}{\|f\|} |Jf| \leqslant \|J\|,$$

i.e., $|Jf| \leqslant \|J\| \, \|f\|$, an inequality which obviously continues to hold if $\|f\| = 0$.

This contradiction shows that $\mu(E) = 0$. By construction, the two continuous linear functionals Jf and

$$I_\varphi f = \int_X f(x)\varphi(x)\,dx$$

coincide on the characteristic functions of all summable sets. But linear combinations of such functions are dense in $L(X)$, and hence $Jf = I_\varphi f$ for all $f \in L(X)$, which is just the relation (10). Moreover,

$$\operatorname*{ess\,sup}_{x\in X} |\varphi(x)| \leqslant \|J\| = \|I_\varphi\| \leqslant \operatorname*{ess\,sup}_{x\in X} |\varphi(x)|,$$

which implies (11) and completes the proof.

9.3.2. The general continuous linear functional on the space $L_p(X)$. Given a set X, let $L_p(X)$, $p > 1$ be the space of all functions $f(x)$ measurable on X such that

$$I(|f|^p) = \int_X |f(x)|^p dx < \infty.$$

Then, as shown in Sec. 6.9, $L_p(X)$ is a complete normed linear space, when equipped with the norm

$$\|f\|_p = I^{1/p}(|f|^p).$$

Let $\varphi(x)$ belong to $L_q(X)$, where

$$\frac{1}{p} + \frac{1}{q} = 1.$$

Then the integral

$$I_\varphi f = I(f\varphi) = \int_X f(x)\varphi(x)\,dx$$

exists and defines a continuous linear functional on $L_p(X)$, since it satisfies the following two conditions:

1) If f_1, f_2 are any two functions in $L_p(X)$ and α_1, α_2 any two real numbers, then

$$I_\varphi(\alpha_1 f_1 + \alpha_2 f_2) = \alpha_1 I_\varphi f_1 + \alpha_2 I_\varphi f_2.$$

2) If $f_m \in L_p(X)$ is a sequence such that $\|f_m\|_p \to 0$ as $m \to \infty$, then $I_\varphi f_m \to 0$, as follows at once from the estimate

$$|I_\varphi f_m| = \left|\int_X f_m(x)\varphi(x)\,dx\right| \leqslant \|f_m\|_p \|\varphi\|_q$$

(use Hölder's inequality, p. 127).

The appropriate analogue of Theorem 4 is now

THEOREM 5. *Given any continuous linear functional Jf defined on the space $L_p(X)$, there exists a function $\varphi(x) \in L_q(X)$ such that*

$$Jf = \int_X f(x)\varphi(x)\,dx, \tag{13}$$

and moreover

$$\|J\| = \|\varphi\|_q. \tag{14}$$

Proof. First suppose $\mu(X) < \infty$, in which case $L_p(X) \subset L(X)$, since

$$|I(|f|)| \leqslant I(|f|^p)I(1^q).$$

For every measurable (and hence summable) set E, with characteristic function χ_E, we write

$$\Phi(E) = J\chi_E.$$

Then

$$|\Phi(E)| = |J\chi_E| \leqslant \|J\| \, \|\chi_E\| = \|J\| \, I^{1/p}(\chi_E^p) = \|J\| \, \mu(E),$$

and hence the set function $\Phi(E)$ is absolutely continuous. Therefore, according to the Radon-Nikodým theorem,

$$\Phi(E) = \int_E \varphi(x) \, dx = I(\chi_E \varphi),$$

where $\varphi(x)$ is summable. Moreover, $\varphi(x)$ belongs to $L_q(X)$ and satisfies the inequality

$$\|\varphi\|_q \leqslant \|J\|.$$

In fact, suppose $0 \leqslant f \leqslant |\varphi|$, where f is bounded and measurable. As on p. 121, the function $f^{q-1} \operatorname{sgn} \varphi$ can be represented as the limit of a uniformly convergent sequence h_n of elementary functions,[5] which we take to be linear combinations of characteristic functions of the measurable sets. It follows that

$$I(f^q) \leqslant I(f^{q-1} |\varphi|) = I(f^{q-1} \operatorname{sgn} \varphi \cdot \varphi) = \lim_{n\to\infty} I(h_n \varphi) = \lim_{n\to\infty} Jh_n$$
$$= J(f^{q-1} \operatorname{sgn} \varphi) \leqslant \|J\| \, \|f^{q-1}\|_p = \|J\| \, I^{1/p}(f^q)$$

(note that h_n also approaches $f^{q-1} \operatorname{sgn} \varphi$ in the L_p-norm). Therefore

$$I^{1-(1/p)}(f^q) = I^{1/q}(f^q) \leqslant \|J\|,$$

and letting $f \nearrow |\varphi|$ (cf. p. 115), we deduce that

$$I^{1/q}(|\varphi|^q) \leqslant \|J\|,$$

i.e., $\varphi \in L_q(X)$ and $\|\varphi\|_q \leqslant \|J\|$, as asserted.

By construction, the two continuous linear functionals Jf and

$$I_\varphi f = \int_X f(x)\varphi(x) \, dx$$

coincide on the characteristic functions of all measurable sets. But linear combinations of such functions are dense in $L_p(X)$ [cf. Theorem 8, p. 130], and hence $Jf = I_\varphi f$ for all $f \in L_p(X)$, which is just the relation (13). Moreover,

$$\|\varphi\|_q \leqslant \|J\| = \|I_\varphi\| \leqslant \|\varphi\|_q, \tag{15}$$

which implies (14).

---

[5] By sgn x is meant the function equal to 1 if $x > 0$, 0 if $x = 0$ and -1 if $x < 0$.

Finally suppose $\mu(X) = \infty$. Then, as in Remark 2, p. 121, $X = X_1 \cup X_2 \cup \cdots$, where $X_1 \subset X_2 \subset \cdots$ and every $\mu(X_n) < \infty$. Given any $f \in L_p(X)$, let

$$f_n(x) = \begin{cases} f(x) & \text{for } x \in X_n, \\ 0 & \text{for } x \in X - X_n. \end{cases}$$

Then obviously $f_n \in L_p(X_n) \subset L_p(X)$, and hence, as just shown, there is a function $\varphi_n \in L_q(X_n) \subset L_q(X)$ such that

$$Jf_n = \int_{X_n} f_n(x)\varphi_n(x)\,dx, \qquad \|\varphi_n\|_q \leqslant \|J\|_n \leqslant \|J\|, \tag{16}$$

where $\|J\|_n$ denotes the norm of J relative to the space $L_p(X_n)$ and $\|J\|$ its norm relative to $L_p(X)$, as before. Let

$$\hat{\varphi}_n(x) = \begin{cases} \varphi_n(x) & \text{for } x \in X_n, \\ 0 & \text{for } x \in X - X_n, \end{cases}$$

and consider the function

$$\varphi(x) = \lim_{n\to\infty} \hat{\varphi}_n(x),$$

where the limit exists, since $\varphi_n(x) = \varphi_{n+1}(x) = \cdots$ if $x \in X_n$. Then (16) can be written in the form

$$Jf_n = \int_X f_n(x)\varphi(x)\,dx, \qquad \int_{X_n} |\varphi(x)|^q\,dx \leqslant \|J\|^q.$$

It follows from the last assertion of Property c, p. 124 that $|\varphi|^q$ is summable on X and

$$\int_X |\varphi(x)|^q\,dx = \lim_{n\to\infty} \int_{X_n} |\varphi(x)|^q\,dx \leqslant \|J\|^q,$$

i.e., $\varphi \in L_q(X)$ and

$$\|\varphi\|_q \leqslant \|J\|.$$

Moreover, by the same argument, $f_n \to f$ in the L_p-norm, and hence

$$Jf = \lim_{n\to\infty} Jf_n = \lim_{n\to\infty} \int_X f_n(x)\varphi(x)\,dx = \int_X f(x)\varphi(x)\,dx.$$

This proves (13), and (14) again follows from (15).

9.4. Positive, Negative and Total Variations of the Sum of Two Set Functions

Let $\Phi_1(E)$ and $\Phi_2(E)$ be two countably additive set functions with positive variations $P_1(E)$ and $P_2(E)$, and let $P(E)$ be the positive variation of their sum

$$\Phi(E) = \Phi_1(E) + \Phi_2(E).$$

For any $A \subset E$ we have

$$\Phi(A) = \Phi_1(A) + \Phi_2(A) \leqslant P_1(E) + P_2(E),$$

and hence

$$P(E) \leqslant P_1(E) + P_2(E), \tag{17}$$

after taking the least upper bound of the left-hand side. The equality

$$P(E) = P_1(E) + P_2(E) \tag{18}$$

does not hold in general, as shown by the example $\Phi_2(E) \equiv -\Phi_1(E) \not\equiv 0$. On the other hand, we can easily establish the following

THEOREM 6. *If $\Phi_1(E)$ and $\Phi_2(E)$ are concentrated on disjoint sets E_1 and E_2, then* (18) *holds.*

Proof. Given any E and $\varepsilon > 0$, we can find sets $A_1 \subset EE_1$ and $A_2 \subset EE_2$ such that

$$P_1(E) = P_1(EE_1) < \Phi_1(A_1) + \varepsilon,$$
$$P_2(E) = P_2(EE_2) < \Phi_2(A_2) + \varepsilon,$$

and hence

$$P(E) \geqslant \Phi(A_1 + A_2) = \Phi(A_1) + \Phi(A_2)$$
$$= \Phi_1(A_1) + \Phi(A_2) > P_1(E) + P_2(E) - 2\varepsilon,$$

since A_1 and A_2 are disjoint. Letting ε approach zero, we obtain the inequality

$$P(E) \geqslant P_1(E) + P_2(E),$$

which, together with (17), implies (18).

COROLLARY 1. *If $\Phi_1(E)$ and $\Phi_2(E)$ are concentrated on disjoint sets E_1 and E_2, then*

$$Q(E) = Q_1(E) + Q_2(E),$$
$$V(E) = V_1(E) + V_2(E),$$

where Q_1, Q_2, Q are the negative variations of the functions $\Phi_1, \Phi_2, \Phi = \Phi_1 + \Phi_2$, and V_1, V_2, V are their total variations.

COROLLARY 2. *The total variation of any countably additive set function equals the sum of the total variation of its continuous component and the total variation of its discrete component.*

COROLLARY 3. *The total variation of any continuous countably additive set function equals the sum of the total variation of its absolutely continuous component and the total variation of its singular component.*

Proof. If $\Phi_1(E)$ is absolutely continuous and $\Phi_2(E)$ is singular (and hence concentrated on a set Z of measure zero), then $\Phi_1(E)$ is concentrated on $X - Z$, since it vanishes on every Borel subset

$$E \subset X - (X - Z) = Z.$$

9.5. The Case $X = [a, b]$. Absolutely Continuous Point Functions

Suppose the countably additive set function $\Phi(E)$ is defined on a σ-ring $\mathfrak{A} = \mathfrak{A}[a, b]$ of subsets of a fixed finite interval $[a, b]$, where $\mathfrak{A}$ contains all subintervals $(\alpha, \beta] \subset [a, b]$,[6] and hence all classical Borel sets. Let $\mu(E)$ be ordinary one-dimensional Lebesgue measure. Then $\Phi(E)$, regarded as a quasi-length, is characterized by its generating function $F(x) = \Phi[a, x]$. Obviously, $F(x)$ is of bounded variation and continuous from the right. Moreover, if $\Phi(E)$ is continuous, then so is $F(x)$, and conversely, since

$$F(x) - F(x - 0) = \lim_{\xi \nearrow x} [F(x) - F(\xi)] = \lim_{\xi \nearrow x} \Phi(\xi, x] = \Phi(\{x\}) = 0.$$

To characterize the generating function of an absolutely continuous set function, we will need the following

DEFINITION. *A point function $F(x)$ defined in the interval $[a, b]$ is said to be absolutely continuous in $[a, b]$ if, given any $\varepsilon > 0$, there exists a $\delta > 0$ such that*

$$\sum_{k=1}^{n} |F(\beta_k) - F(\alpha_k)| < \varepsilon \tag{19}$$

for every collection of disjoint subintervals $(\alpha_k, \beta_k] \subset [a, b]$ such that

$$\sum_{k=1}^{n} (\beta_k - \alpha_k) < \delta.$$

Remark. An equivalent definition is obtained by replacing the inequality (19) by

$$\left| \sum_{k=1}^{n} [F(\beta_k) - F(\alpha_k)] \right| < \varepsilon$$

[obviously implied by (19)]. In fact, given any $\varepsilon > 0$, let $\delta > 0$ be such that

$$\left| \sum_{k=1}^{n} [F(\beta_k) - F(\alpha_k)] \right| < \frac{\varepsilon}{2} \tag{20}$$

for every collection of disjoint subintervals $(\alpha_k, \beta_k] \subset [a, b]$ whose total length is less than δ. In general, $F(\beta_k) - F(\alpha_k) \geqslant 0$ for some of the subintervals $(\alpha_k, \beta_k]$, and $F(\beta_k) - F(\alpha_k) < 0$ for the others. Correspondingly, the sum in (20) splits into two sums

$$\sum_{k=1}^{n}{}^{+} [F(\beta_k) - F(\alpha_k)], \qquad \sum_{k=1}^{n}{}^{-} [F(\beta_k) - F(\alpha_k)],$$

[6] As usual, we replace $(\alpha, \beta]$ by $[\alpha, \beta]$ if $\alpha = a$. Note that $\Phi(E)$ is defined for $E = X = [a, b]$.

where

$$\sum_{k=1}^{n}{}^{+}[F(\beta_k) - F(\alpha_k)] < \frac{\varepsilon}{2}, \qquad \left|\sum_{k=1}^{n}{}^{-}[F(\beta_k) - F(\alpha_k)]\right| < \frac{\varepsilon}{2},$$

since the sum of the lengths of each collection of subintervals is obviously less than δ. But then

$$\sum_{k=1}^{n} |F(\beta_k) - F(\alpha_k)| = \sum_{k=1}^{n}{}^{+}[F(\beta_k) - F(\alpha_k)] + \left|\sum_{k=1}^{n}{}^{-}[F(\beta_k) - F(\alpha_k)]\right| < \varepsilon,$$

i.e., (19) holds, as required.

Next we note some simple properties of absolutely continuous point functions:

1) *If $F(x)$ is absolutely continuous in $[a, b]$, then $F(x)$ is uniformly continuous in $[a, b]$.*

2) *If $F(x)$ is absolutely continuous in $[a, b]$, then $F(x)$ is of bounded variation in $[a, b]$.* In fact, if δ corresponds to the choice $\varepsilon = 1$ in (19), then the total variation of $F(x)$ is less than 1 in every subinterval $(\alpha_k, \beta_k] \subset [a, b]$ of length less than δ. Therefore

$$V_a^b(F) < N \cdot 1 < \frac{b-a}{\delta} + 1,$$

where N is the smallest number of disjoint subintervals $(\alpha_k, \beta_k]$ needed to cover $[a, b]$ and $V_a^b(F)$ is the total variation of $F(x)$ in $[a, b]$.

3) *If $F(x)$ is absolutely continuous in $[a, b]$, then so are its positive, negative and total variations.* For example, let $V(x)$ be the total variation of $F(x)$, and given any $\varepsilon > 0$, let $\delta > 0$ be such that the inequality (19) holds for every collection of disjoint subintervals $(\alpha_k, \beta_k]$ with total length less than δ. By definition of the total variation (cf. p. 85), the sum

$$\sum_{k=1}^{n} [V(\beta_k) - V(\alpha_k)] = \sum_{k=1}^{n} V(\alpha_k, \beta_k] \tag{21}$$

is the least upper bound of the quantity

$$\sum_{k=1}^{N} \sum_{j=1}^{m_k} |F(x_j^{(k)}) - F(x_{j-1}^{(k)})|, \tag{22}$$

where $\alpha_k = x_0^{(k)} < \cdots < x_{m_k}^{(k)} = \beta_k$ is an arbitrary partition of the interval $(\alpha_k, \beta_k]$. By construction, the sum of the lengths of the disjoint subintervals $(x_{j-1}^{(k)}, x_j^{(k)}]$ is less than δ, and hence every sum (22) is less than ε. But then their least upper bound (21) cannot exceed ε, i.e., $V(x)$ is itself absolutely continuous, as asserted.

After these preliminaries, we now prove

THEOREM 7. *A countably additive set function* $\Phi(E)$ *defined on* $\mathfrak{A} = \mathfrak{A}[a, b]$ *is absolutely continuous if and only if its generating function* $F(x)$ *is absolutely continuous in* $[a, b]$.

Proof. Let $F(x)$ be absolutely continuous in $[a, b]$, where, by Property 3 above, there is no loss of generality in assuming that $F(x)$ is nondecreasing. Given any $\varepsilon > 0$, choose $\delta > 0$ such that

$$\sum_{k=1}^{n} [F(\beta_k) - F(\alpha_k)] < \varepsilon \tag{23}$$

for every disjoint collection of subintervals $(\alpha_k, \beta_k] \subset [a, b]$ of total length less than δ. Then, given any $Z \in \mathfrak{A}$ of measure zero, let $\{\Delta_k\}$ be a countable collection of subintervals of total length less than δ covering Z (cf. Sec. 1.4). These intervals can be made disjoint by the usual device of replacing $\Delta_1, \Delta_2, \Delta_3, \ldots$ by

$$\Delta_1, \Delta_1 - \Delta_1\Delta_2, \Delta_3 - \Delta_1\Delta_3 - \Delta_2\Delta_3, \ldots,$$

if necessary. Because of (23), we have

$$\Phi\left(\bigcup_{k=1}^{n} \Delta_k\right) = \sum_{k=1}^{n} \Phi(\Delta_k) < \varepsilon$$

for every finite subcollection $\Delta_1, \ldots, \Delta_n$. But $\Phi(E)$ is countably additive, and hence

$$\Phi\left(\bigcup_{k=1}^{\infty} \Delta_k\right) = \lim_{n\to\infty} \Phi\left(\bigcup_{k=1}^{n} \Delta_k\right) \leqslant \varepsilon.$$

which implies

$$\Phi(Z) \leqslant \varepsilon,$$

since

$$Z \subset \bigcup_{k=1}^{\infty} \Delta_k.$$

Therefore $\Phi(Z) = 0$, since $\varepsilon > 0$ is arbitrary.

Conversely, if $\Phi(E)$ is absolutely continuous, then, by the Radon-Nikodým theorem,

$$\Phi(E) = \int_E g(x)\, dx,$$

where $g(x)$ is summable on $[a, b]$. In particular,

$$F(x) = \Phi[a, x] = \int_a^x g(\xi)\, d\xi,$$

and hence

$$\sum_{k=1}^{n} [F(\beta_k) - F(\alpha_k)] = \sum_{k=1}^{n} \int_{\alpha_k}^{\beta_k} g(\xi)\, d\xi = \int_{\Delta_1 \cup \cdots \cup \Delta_n} g(\xi)\, d\xi, \tag{24}$$

where $\Delta_k = (\alpha_k, \beta_k]$. But by Property d, p. 124, the right-hand side of (24) approaches zero as the total length of the intervals $\Delta_1, \ldots, \Delta_n$ goes to zero. This completes the proof (recall the remark on p. 196).

COROLLARY. *If $F(x)$ is absolutely continuous in $[a, b]$, then*

$$F(x) = F(a) + \int_a^x g(\xi)\, d\xi, \tag{25}$$

where $g(x)$ is summable on $[a, b]$. Conversely, every function of the form (25) is absolutely continuous in $[a, b]$.

9.6. The Lebesgue Decomposition

We continue our study of the case $X = [a, b]$, by allowing $\Phi(E)$ to have singular and discrete components. First we prove

THEOREM 8. *A countably additive set function $\Phi(E)$ defined on $\mathfrak{A} = \mathfrak{A}[a, b]$ is singular if and only if its generating function $F(x)$ is singular in $[a, b]$ in the following sense:*[7] *Given any $\varepsilon > 0$, there exists a finite collection of subintervals $(\alpha_k, \beta_k] \subset [a, b]$ such that*

$$\sum_{k=1}^{n}(\beta_k - \alpha_k) < \varepsilon, \qquad \sum_{k=1}^{n} |F(\beta_k) - F(\alpha_k)| > V_a^b(F) - \varepsilon. \tag{26}$$

Proof. First suppose $\Phi(E)$ is singular, and let Z be the set of measure zero on which $\Phi(E)$, and hence its total variation $V(E)$, is concentrated. As we know from Sec. 4.7.4, regarded as a quasi-length, $V(E)$ has the generating function

$$V_a^x(F) = \sup \sum_{j=1}^{m} |F(x_j) - F(x_{j-1})| ,$$

where the least upper bound is taken with respect to all partitions $a = x_0 < \cdots < x_m = x$ of the interval $[a, x]$.[8] Given any $\varepsilon > 0$, let $(\alpha_k, \beta_k]$, $k = 1, 2, \ldots$ be a countable collection of disjoint subintervals covering Z such that

$$\sum_{k=1}^{\infty}(\beta_k - \alpha_k) < \varepsilon.$$

[7] This presupposes that a singular function is of bounded variation. A condition equivalent to (26) is the following: Given any $\varepsilon > 0$, there exists a finite collection of subintervals $(\alpha_k', \beta_k') \subset [a, b]$ such that

$$\sum_{k=1}^{n}(\beta_k' - \alpha_k') > b - a - \varepsilon, \qquad \sum_{k=1}^{n} |F(\beta_k') - F(\alpha_k')| < \varepsilon.$$

[8] In particular, $V_a^b(F) = V[a, b]$.

Clearly

$$\sum_{k=1}^{\infty} V(\alpha_k, \beta_k] = V_a^b(F),$$

since $V(E)$ is concentrated on Z. Let $n = n(\varepsilon)$ be such that

$$\sum_{k=1}^{n} V(\alpha_k, \beta_k] > V_a^b(F) - \frac{\varepsilon}{2},$$

and then, in each of the n subintervals $(\alpha_1, \beta_1], \ldots, (\alpha_n, \beta_n]$, choose points $\alpha_k = x_0^{(k)} < \cdots < x_{m_k}^{(k)} = \beta_k$ such that

$$\sum_{j=1}^{m_k} |F(x_j^{(k)}) - F(x_{j-1}^{(k)})| > V(\alpha_k, \beta_k] - \frac{\varepsilon}{2n}.$$

It follows that

$$\sum_{k=1}^{n} \sum_{j=1}^{m_k} |F(x_j^{(k)}) - F(x_{j-1}^{(k)})| > \sum_{k=1}^{n} V(\alpha_k, \beta_k] - \frac{\varepsilon}{2} > V_a^b(F) - \varepsilon,$$

which proves (26), since obviously

$$\sum_{k=1}^{n} (\beta_k - \alpha_k) < \varepsilon.$$

Conversely, suppose $F(x)$ satisfies (26). Then, by Theorems 1 and 3, $\Phi(E)$ can be written in the form

$$\Phi(E) = A(E) + S(E),$$

where $A(E)$ is absolutely continuous and $S(E)$ is singular but not necessarily continuous.[9] Correspondingly, $F(x)$ itself has the representation

$$F(x) = A(x) + S(x), \tag{27}$$

where, by Theorem 7, $A(x)$ is absolutely continuous. By hypothesis, given any $\varepsilon > 0$, there is a collection of disjoint subintervals $(\alpha_k, \beta_k] \subset [a, b]$ of total length less than ε such that

$$\sum_{k=1}^{n} |F(\beta_k) - F(\alpha_k)| > V_a^b(F) - \varepsilon,$$

and hence

$$\sum_{k=1}^{n} |S(\beta_k) - S(\alpha_k)| > V_a^b(F) - \sum_{k=1}^{n} |A(\beta_k) - A(\alpha_k)| - \varepsilon,$$

because of (27). The sum on the right approaches zero as $\varepsilon \to 0$, because of the absolute continuity of $A(x)$. Therefore

$$\sum_{k=1}^{n} |S(\beta_k) - S(\alpha_k)| > V_a^b(F) - \varepsilon',$$

---

[9] Recall from p. 184 that a discrete set function is singular.

where $\varepsilon' > 0$ is arbitrary, and hence the total variation of $S(x)$ cannot be less than $V_a^b(F)$. But then, by Theorem 6, Corollary 3, the total variation of $A(x)$ vanishes. It follows that $A(x) \equiv 0$, and hence

$$F(x) = S(x), \qquad \Phi(E) = S(E),$$

as required.

Next we consider the case where $\Phi(E)$ is discrete. Then, according to Sec. 9.1,

$$\Phi(E) = \sum_{c_k \in E} g_k, \tag{28}$$

where $c_1, c_2, \ldots$ is a sequence of points in $[a, b]$ and $g_1, g_2, \ldots$ is a corresponding sequence of real numbers such that

$$\sum_{k=1}^{\infty} |g_k| < \infty.$$

In particular, the generating function of $\Phi(E)$ is given by

$$F(x) = \sum_{c_k \in [a, x]} g_k.$$

THEOREM 9. *A countably additive set function $\Phi(E)$ defined on $\mathfrak{A} = \mathfrak{A}[a, b]$ is discrete if and only if its generating function $F(x)$ is a jump function in the following sense: $F(x)$ is continuous from the right in $[a, b]$,*[10] *and given any $\varepsilon > 0$, there exist finitely many points of discontinuity $c_1, \ldots, c_n$ of $F(x)$ such that*[11]

$$\sum_{k=1}^{n} |F(c_k) - F(c_k - 0)| > V_a^b(F) - \varepsilon. \tag{29}$$

Proof. First suppose $\Phi(E)$ is discrete, and let $c_1, c_2, \ldots$ and $g_1, g_2, \ldots$ be the points and real numbers figuring in (28). Then

$$|F(c_k) - F(c_k - 0)| = |\Phi(\{c_k\})| = |g_k|,$$

$$V_a^b(F) = \sum_{k=1}^{\infty} |g_k|,$$

which implies (29). Conversely, suppose $F(x)$ satisfies (29). Then, by Theorem 1,

$$\Phi(E) = C(E) + D(E),$$

where $C(E)$ is continuous and $D(E)$ is discrete. Correspondingly, $F(x)$ has the representation

$$F(x) = C(x) + D(x),$$

---

[10] Recall that the generating function of a countably additive set function $\Phi(E)$ is automatically continuous from the right, since $\Phi(E)$ is (upper) continuous, regarded as a quasi-volume (cf. Sec. 5.6).

[11] In particular, $F(x)$ can have no more than countably many points of discontinuity.

where $C(x)$ is continuous. By hypothesis, given any $\varepsilon > 0$, there are points $c_1, \ldots, c_n \in [a, b]$ such that

$$\sum_{k=1}^{n} |F(c_k) - F(c_k - 0)| > V_a^b(F) - \varepsilon,$$

or equivalently,

$$\sum_{k=1}^{n} |D(c_k) - D(c_k - 0)| > V_a^b(F) - \varepsilon,$$

since $C(x)$ is continuous. Therefore the total variation of $D(x)$ cannot be less than $V_a^b(F)$. But then, by Theorem 6, Corollary 2, the total variation of $C(x)$ vanishes. It follows that $C(x) \equiv 0$, and hence

$$F(x) = D(x), \qquad \Phi(E) = D(E),$$

as required.

Finally, combining Theorem 3 and the last three theorems, we obtain

THEOREM 10 (*Lebesgue decomposition*). *If $F(x)$ is the generating function of a countably additive set function $\Phi(E)$ defined on $\mathfrak{A} = \mathfrak{A}[a, b]$, then $F(x)$ can be represented in the form*

$$F(x) = A(x) + S(x) + D(x), \tag{30}$$

where $A(x)$ is absolutely continuous, $S(x)$ is continuous and singular, and $D(x)$ is a jump function. Moreover, $A(x)$ itself has the representation

$$A(x) = \int_a^x g(\xi)\, d\xi,$$

where the function $g(x)$ is summable on $[a, b]$. The representation (30) is unique, and so is $g(x)$, to within a set of measure zero.

COROLLARY 1. *The conclusion of the theorem remains true if $F(x)$ is of bounded variation and continuous from the right in $[a, b]$, except that now*

$$A(x) = F(a) + \int_a^x g(\xi)\, d\xi.$$

Proof. Consider the generating function $F(x) - F(a)$.

COROLLARY 2. *The conclusion of the theorem remains true if $F(x)$ is of bounded variation in $[a, b]$, but not necessarily continuous from the right, except that now*

$$A(x) = F(a) + \int_a^x g(\xi)\, d\xi$$

and $D(x)$ is a jump function in the sense that given any $\varepsilon > 0$, there exist finitely many points of discontinuity $c_1, \ldots, c_n$ of $D(x)$ such that[12]

$$\sum_{k=1}^{n} |D(c_k + 0) - D(c_k - 0)| > V_a^b(D) - \varepsilon.$$

[12] Of course, in general, $D(x)$ itself is no longer continuous from the right. Again, $D(x)$ can have no more than countably many points of discontinuity.

Proof. As we know from p. 85, $F(x)$ is the difference between two bounded nonnegative nondecreasing functions. Therefore, by an elementary argument,[13] $F(x - 0)$ and $F(x + 0)$ exist for all $x \in [a, b]$. Moreover, there are at most countably many points where $F(x - 0) \neq F(x + 0)$, since the inequality

$$|F(x + 0) - F(x - 0)| > \frac{1}{n} \qquad (n = 1, 2, \ldots)$$

can hold at no more than $N(n)$ points, where $N(n)$ is the largest integer in

$$\frac{1}{n} V_a^b(F).$$

Hence there is a jump function $D_1(x)$ such that $F(x) + D_1(x)$ is continuous from the right. The proof now follows by applying Corollary 1 to the function $F(x) + D_1(x)$.

PROBLEMS

1. Let $\mathscr{A}$ be the space of all absolutely continuous functions defined on the closed interval $[a, b]$. Show that $\mathscr{A}$ is a closed subspace of the space $\mathscr{B}$ of all functions of bounded variation in $[a, b]$ (cf. Prob. 8, p. 87).

2. Let $\mathscr{S}$ be the space of all singular functions defined on $[a, b]$. Show that $\mathscr{S}$ is a closed subspace of $\mathscr{B}$.

3. Let $\mathscr{D}$ be the space of all jump functions defined on $[a, b]$. Show that $\mathscr{D}$ is a closed subspace of $\mathscr{B}$.

4. Show that the Cantor function $C(x)$ of Prob. 2, p. 86 is singular.

Hint. The corresponding set function $C(E)$ vanishes in every interval adjacent to the Cantor set (see Prob. 2, p. 21). Recall footnote 7, p. 199.

5. Show that if every term $\Phi(E_1), \Phi(E_2), \ldots$ of an everywhere convergent series

$$\Phi(E) = \Phi(E_1) + \Phi(E_2) + \cdots$$

is a nonnegative countably additive set function defined on some σ-ring $\mathfrak{A}$, then so is the sum $\Phi(E)$. Show that if every term is absolutely continuous (relative to some measure μ), then so is $\Phi(E)$.

Hint. The relation

$$\sum_{m=1}^{n} \Phi(E_m) = \lim_{p\to\infty} \sum_{m=1}^{n} \sum_{k=1}^{p} \Phi_k(E_m) \leqslant \lim_{p\to\infty} \Phi_k(E) = \Phi(E)$$

---

[13] See e.g., T. M. Apostol, *Mathematical Analysis*, Addison-Wesley Publishing Co., Reading, Mass. (1957), p. 78.

implies

$$\sum_{m=1}^{\infty} \Phi(E_m) \leqslant \Phi(E).$$

On the other hand, given any $\varepsilon > 0$, we can choose p such that

$$\sum_{k=1}^{p} \Phi_k(E) > \Phi(E) - \varepsilon$$

and then n such that

$$\sum_{m=1}^{n} \Phi_k(E_m) > \Phi_k(E) - \frac{\varepsilon}{p} \qquad (k = 1, \ldots, p).$$

Therefore

$$\sum_{m=1}^{n} \Phi(E_m) \geqslant \sum_{k=1}^{p} \sum_{m=1}^{n} \Phi_k(E_m) \geqslant \sum_{k=1}^{p} \Phi_k(E) - \varepsilon > \Phi(E) - 2\varepsilon,$$

and hence

$$\sum_{m=1}^{\infty} \Phi(E_m) \geqslant \Phi(E).$$

6. Construct a singular function $F(x)$ on the interval $[0, 1]$ with no intervals of constancy.

Hint. Construct $F(x)$ as a series of functions of the Cantor type (cf. Prob. 4) such that $F(x)$ is concentrated on a set Z dense in $[0, 1]$. Use one of the results of Prob. 5.

7. Using the function $F(x)$ of Prob. 6, show that a measurable function of a continuous function need not be measurable.

Hint. The function $F(x)$ maps a set of full measure into a set of measure zero. Consider a nonmeasurable subset $W \subset E$ and let $G(y)$ be the characteristic function of the set $F(W)$. Show that $G[F(y)]$ is nonmeasurable.

8. Prove that a nondecreasing function $F(x) \not\equiv$ const is singular if and only if it maps some set of measure zero into a set of full measure, or some set of full measure into a set of measure zero.

9. Prove that the inverse of a continuous singular function with no intervals of constancy is itself singular.

Hint. Use Prob. 8.

10. Let $\Phi(E)$ be a countably additive set function defined on a σ-ring $\mathfrak{A}$ of subsets of X, equipped with a Borel measure μ, and suppose $\Phi(E)$ is not absolutely continuous with respect to μ. Let $\overline{\mathfrak{A}}$ be a Lebesgue extension of $\mathfrak{A}$. Show that in general $\Phi(E)$ cannot be extended onto the σ_μ-ring $\overline{\mathfrak{A}}$.

Hint. Let X be the interval $[a, b]$, let μ be ordinary Lebesgue measure, and let $\Phi(E)$ be a countably additive set function defined on all the Borel subsets of X. If $\Phi(E)$ is continuous but not absolutely continuous, there exists a noncountable set $E_0 \subset X$, $\mu(E_0) = 0$ such that $\Phi(E_0) \neq 0$. Construct a Φ-nonmeasurable set $E_1 \subset E_0$ which is Lebesgue measurable (with measure zero).

10

THE DERIVATIVE OF A SET FUNCTION

10.1. Preliminaries. Various Definitions of the Derivative

Let $\Phi(E)$ be a countably additive (finite) set function defined on a σ-ring $\mathfrak{A}$ of subsets $E \subset X$, equipped with a nonnegative Borel measure $\mu(E)$. As in Sec. 9.1, if $X \notin \mathfrak{A}$, we assume that $X = X_1 \cup X_2 \cup \cdots$, where $X_1 \subset X_2 \subset \cdots$ and every $X_n \in \mathfrak{A}$. This situation will be summarized by saying that "X is a set equipped with a measure μ" and "$\Phi(E)$ is a countably additive set function (defined) on X."

Suppose $\Phi(E)$ is absolutely continuous with respect to $\mu(E)$. Then, according to the Radon-Nikodým theorem, $\Phi(E)$ can be represented in the form

$$\Phi(E) = \int_E g(x)\mu(dx)$$

in terms of a μ-summable function $g(x)$, which we shall call the *density* of $\Phi(E)$. This immediately raises the question of how to find the density of $\Phi(E)$, starting from a knowledge of $\Phi(E)$ itself.

In simple cases, the procedure for finding $g(x)$ is familiar from elementary calculus. For example, let X be a finite interval $[a, b]$, equipped with ordinary Lebesgue measure, and let $\Phi(E)$ be the set function with generating function $F(x)$, so that

$$F(x) = \Phi[a, x] = \int_a^x g(\xi)\, d\xi.$$

If $g(x)$ is continuous, then, as is well known, $g(x)$ can be obtained by differentiating $F(x)$, i.e.,

$$g(x_0) = \lim_{h \to 0} \frac{F(x_0 + h) - F(x_0)}{h} = \lim_{h \to 0} \frac{\Phi(x_0, x_0 + h]}{h}, \tag{1}$$

where $\Phi(x_0, x_0 + h] = -\Phi(x_0 + h, x_0]$ if $h < 0$. Similarly, if $\mathbf{B}$ is an n-dimensional block, again equipped with ordinary Lebesgue measure, and if $g(x) = g(x_1, \ldots, x_n)$ is continuous, then

$$g(x_0) = \lim_{B \to x_0} \frac{1}{s(B)} \int_B g(\xi)\, d\xi,$$

where $s(B)$ is the volume of the block $B \subset \mathbf{B}$, and the limit is taken with respect to an arbitrary sequence of blocks $\{B_n\}$ converging to the point x_0.[1]

Formula (1) gives the usual definition of differentiation with respect to one-dimensional Lebesgue measure, but it is hardly the unique definition. In fact, as we now show, (1) can be either weakened or strengthened, while remaining a perfectly plausible definition.

Suppose first that instead of using intervals of the form $(x, x + h]$, we restrict ourselves to intervals from a much smaller class, for example, intervals of the form

$$\left(\frac{p}{2^n}, \frac{p+1}{2^n}\right],$$

whose end points are binary numbers, where every interval contains the point x_0 at which the derivative is to be evaluated. In other words, we define the derivative of the countably additive set function $\Phi(E)$ by the formula

$$F'(x_0) = \lim_{n \to \infty} \frac{\Phi\left(\dfrac{p}{2^n}, \dfrac{p+1}{2^n}\right]}{\dfrac{1}{2^n}} \equiv \lim_{n \to \infty} \frac{F\left(\dfrac{p+1}{2^n}\right) - F\left(\dfrac{p}{2^n}\right)}{\dfrac{1}{2^n}}, \tag{2}$$

$$x_0 \in \left(\frac{p}{2^n}, \frac{p+1}{2^n}\right],$$

provided the limit exists. This definition is weaker than (1), in the sense that the existence of the limit (1) implies that of (2), but not conversely. In fact, suppose the limit (1) exists at some point x_0. Then the expression

$$\frac{\Phi\left(\dfrac{p}{2^n}, \dfrac{p+1}{2^n}\right]}{\dfrac{1}{2^n}} = \frac{\Phi\left(\dfrac{p}{2^n}, x_0\right]}{\dfrac{1}{2^n}} + \frac{\Phi\left(x_0, \dfrac{p+1}{2^n}\right]}{\dfrac{1}{2^n}}$$

$$= \frac{\Phi\left(\dfrac{p}{2^n}, x_0\right]}{x_0 - \dfrac{p}{2^n}} \frac{x_0 - \dfrac{p}{2^n}}{\dfrac{1}{2^n}} + \frac{\Phi\left(x_0, \dfrac{p+1}{2^n}\right]}{\dfrac{p+1}{2^n} - x_0} \frac{\dfrac{p+1}{2^n} - x_0}{\dfrac{1}{2^n}}$$

[1] In the sense that the size of B_n (see p. 7) approaches zero as $n \to \infty$, and

$$\bigcap_{n=1}^{\infty} B_n = \{x_0\}.$$

is a weighted average of two expressions approaching $F'(x_0)$ as $n \to \infty$, and hence itself approaches $F'(x_0)$. On the other hand, as shown in Prob. 1, p. 223, given an irrational point x_0, we can always construct a continuous function $F(x)$ of bounded variation such that

a) $F(x)$ has no derivative in the ordinary sense at the point x_0;
b) $F(x)$ vanishes at the end points of the intervals $p2^{-n} < x \leqslant (p+1)2^{-n}$ and hence has a (zero) derivative at x_0 in the sense of formula (2).

Next, instead of using intervals of the form $(x, x+h]$, we consider sets from a much *larger* class. Thus we now define the derivative of the countably additive set function $\Phi(E)$ by the formula

$$\lim_{n\to\infty} \frac{\Phi(E_n)}{\mu(E_n)} \tag{3}$$

(provided the limit exists), where μ is ordinary Lebesgue measure and $E_1, E_2, \ldots$ is an arbitrary sequence of Borel sets "converging regularly" to the point x_0 in the sense that

1) Every E_n is contained in a half-open interval $\Delta_n = (a_n, b_n]$ such that $x_0 \in \Delta_n$ and $a_n, b_n \to x_0$ as $n \to \infty$;
2) There is a fixed constant $c > 0$ such that

$$\mu(E_n) \geqslant c\mu(\Delta_n)$$

for every n.

This definition is stronger than (1), since the existence of the limit (3) implies that of (1), but not conversely. In fact, suppose the limit (3) exists at some point x_0, and choose $E_n = (x_0, x_0 + h_n]$, $\Delta_n = (x_0 - h_n, x_0 + h_n]$, where h_n is an arbitrary sequence converging to zero as $n \to \infty$. Then E_n converges regularly to the point x_0 (with $c = \frac{1}{2}$), so that the limit

$$\lim_{n\to\infty} \frac{\Phi(E_n)}{\mu(E_n)} = \lim_{n\to\infty} \frac{\Phi(x_0, x_0 + h_n]}{h_n},$$

equivalent to (1), exists (there are obvious changes for negative h_n). On the other hand, the derivative of the function

$$x^2 \sin \frac{1}{x}$$

in the sense of formula (1) exists and equals zero at the point $x_0 = 0$, whereas the derivative in the sense of formula (3) does not exist (see Prob. 2, p. 223).

The above considerations notwithstanding, it turns out that differences between the definitions (1), (2) and (3) appear only "at separate points." In fact, as we shall soon see, *the three definitions* (1), (2) *and* (3) *of the derivative of a function of bounded variation* (*or their analogues for a countably* *additive set function defined on a general set X) exist and coincide on a set of full measure.*

10.2. Differentiation with Respect to a Net

The appropriate generalization of differentiation in the sense of formula (2) is differentiation with respect to a *net*. Let X be a set equipped with a measure μ, and suppose X is the union of a countable family $\mathfrak{N}_1$ of disjoint Borel sets $A_1^{(1)}, \ldots, A_j^{(1)}, \ldots$, called sets *of the first rank*. Suppose further that every set $A_j^{(1)}$ of the first rank is itself the union of a countable family of disjoint Borel sets $A_{j1}^{(2)}, \ldots, A_{jk}^{(2)}, \ldots$, called sets *of the second rank*, and let $\mathfrak{N}_2$ denote the family of all such sets where j is arbitrary. Imagine this process repeated for every integer n, leading in each case to a family $\mathfrak{N}_n$ of disjoint Borel sets, said to be *of the n'th rank*, whose union is the original set X. Then the family

$$\mathfrak{N} = \bigcup_{n=1}^{\infty} \mathfrak{N}_n$$

of all sets of every finite rank is called a *net*, provided $\mathfrak{N}$ is completely sufficient in the sense of Sec. 7.4 (note that $\mathfrak{N}$ is a semiring).

Now let $\Phi(E)$ be a countably additive set function defined on X (and hence on $\mathfrak{N}$). Then, by the *derivative of $\Phi(E)$ at the point x_0 with respect to the net $\mathfrak{N}$*, we mean the quantity

$$D_{\mathfrak{N}}(x_0) = \lim_{n \to \infty} \frac{\Phi[A_n(x_0)]}{\mu[A_n(x_0)]} \tag{4}$$

(provided the limit exists), where $A_n(x_0)$ is the unique set of the nth rank containing x_0. According to a theorem of de Possel,[2] the derivative of $\Phi(E)$ with respect to any net $\mathfrak{N}$ exists on a set of full measure (for suitable X) and coincides with the density of the absolutely continuous component of $\Phi(E)$. In particular, it follows that the derivative is independent of $\mathfrak{N}$.

Example. Let X be the unit interval $[0, 1]$, equipped with ordinary Lebesgue measure, and let the sets of the nth rank be

$$\left[0, \frac{1}{2^n}\right], \left(\frac{1}{2^n}, \frac{2}{2^n}\right], \ldots, \left(1 - \frac{1}{2^n}, 1\right].$$

Then the derivative with respect to the corresponding net $\mathfrak{N}$ is the derivative in the sense of formula (2).

[2] De Possel's theorem will be deduced in Sec. 10.4 from the more general Lebesgue-Vitali theorem, which treats the analogue of "ordinary differentiation," i.e., differentiation in the sense of formula (1).

10.3. Differentiation with Respect to a Vitali System. The Lebesgue-Vitali Theorem

Next we consider the generalization of differentiation in the sense of formula (1). Again let X be a set equipped with a measure μ, but now suppose every set $\{x\}$ consisting of a single point $x \in X$ is measurable, with measure zero. By a *Vitali system* we mean a family $\mathscr{V}$ of Borel sets $E \subset X$ (called *Vitali sets*) which has the following properties:

1) Given any Borel set (or generalized Borel set) E and any $\varepsilon > 0$, there are countably many Vitali sets $A_1, A_2, \ldots$ such that

$$\bigcup_{n=1}^{\infty} A_n \supset E, \qquad \mu\left(\bigcup_{n=1}^{\infty} A_n\right) < \mu(E) + \varepsilon.$$

2) Every set $E \in \mathscr{V}$ has a *boundary*, i.e., a set $\Gamma(E)$ of measure zero such that

   a) If $x \in E - E\Gamma(E)$, then every Vitali set of sufficiently small measure containing x is contained in $E - E\Gamma(E)$;

   b) If $x \notin \bar{E} = E \cup \Gamma(E)$, then every Vitali set of sufficiently small measure containing x does not intersect $\bar{E}$.

3) Suppose $E \subset X$ is a set covered by a subsystem $\mathfrak{B} \subset \mathscr{V}$ of Vitali sets such that for any $x \in E$ and any $\varepsilon > 0$, there is a set $A_\varepsilon(x) \in \mathfrak{B}$ of measure less than ε which contains x. Then E can be covered to within a set of measure zero by countably many disjoint sets $A_j \in \mathfrak{B}$.

Now let $\Phi(E)$ be a countably additive set function defined on X (and hence on $\mathscr{V}$). *Then, by the derivative of* $\Phi(E)$ *at the point* x_0 *with respect to the Vitali system* $\mathscr{V}$, we mean the quantity

$$D_{\mathscr{V}}\Phi(x_0) \equiv D\Phi(x_0) = \lim_{\varepsilon \to 0} \frac{\Phi[A_\varepsilon(x_0)]}{\mu[A_\varepsilon(x_0)]}$$

(provided the limit exists), where $A_\varepsilon(x_0)$ is any Vitali set of measure less than ε containing the point x_0. In any event, the quantities

$$\bar{D}\Phi(x_0) \equiv \overline{\lim_{\varepsilon \to 0}} \frac{\Phi[A_\varepsilon(x_0)]}{\mu[A_\varepsilon(x_0)]}$$

and

$$\underline{D}\Phi(x_0) \equiv \underline{\lim}_{\varepsilon \to 0} \frac{\Phi[A_\varepsilon(x_0)]}{\mu[A_\varepsilon(x_0)]}$$

always exist (provided we allow the values $\pm\infty$), where $\bar{D}\Phi(x_0)$ is called the *upper derivative* and $\underline{D}\Phi(x_0)$ the *lower derivative* of $\Phi(E)$ at the point x_0 with respect to the Vitali system $\mathscr{V}$. A necessary and sufficient condition for $\Phi(E)$ to have a derivative at x_0 with respect to $\mathscr{V}$ is obviously that

$$\bar{D}\Phi(x_0) = \underline{D}\Phi(x_0) \neq \infty.$$

In general, the quantities $\bar{D}\Phi(x_0)$, $\underline{D}\Phi(x_0)$ and $D\Phi(x_0)$ depend not only on the function $\Phi(E)$ but also on the underlying Vitali system $\mathscr{V}$. However, as we shall see presently, effects of this sort manifest themselves only on a set of measure zero. First we establish some necessary preliminary results:

LEMMA 1. *If the inequality*

$$\bar{D}\Phi(x) = \overline{\lim_{\varepsilon\to 0}} \frac{\Phi[A_\varepsilon(x)]}{\mu[A_\varepsilon(x)]} > c$$

(c fixed) holds at every point x of a set $E \subset X$ of positive measure (where X is a Borel set[3]*), then, given any $\varepsilon > 0$, there is a set $Q \subset E$ such that*

$$\mu(E - Q) < \varepsilon, \qquad \Phi(Q) \geqslant c\mu(Q).$$

Proof. Given any $\varepsilon > 0$, we use Property 1 to cover the complement $\mathscr{C}E$ of the set E by a countable family of Vitali sets $B_1, B_2, \ldots$ such that[4]

$$\mu\left(\bigcup_{n=1}^{\infty} B_n\right) < \mu(\mathscr{C}E) + \varepsilon.$$

Letting B denote the union of all the sets $\bar{B}_n = B_n \cup \Gamma(B_n)$, we now use mathematical induction to construct a sequence of sets $Q_1, Q_2, \ldots$, each a union of no more than countably many disjoint Vitali sets, such that $\Phi(Q_n) > c\mu(Q_n)$ and

$$E - EB - Z_n \subset Q_n \subset \mathscr{C}(\bar{B}_1 \cup \cdots \cup \bar{B}_n),$$

where each Z_n $(n = 1, 2, \ldots)$ is a set of measure zero. First setting $n = 1$, for every point $x \in E - BE$ we find all the Vitali sets $A_\alpha^{(1)}(x)$ which contain x,[5] do not intersect $\bar{B}_1$ and satisfy the inequality

$$\frac{\Phi[A_\alpha^{(1)}(x)]}{\mu[A_\alpha^{(1)}(x)]} > c.$$

By Property 2b such sets exist and have arbitrarily small measure. Then from this covering of $E - BE$, we use Property 3 to select a sequence of disjoint Vitali sets $A_j^{(1)}$ covering $E - BE$ to within a set Z_1 of measure zero. Thus, if Q_1 is the union of all the $A_j^{(1)}$, we have

$$E - BE - Z_1 \subset Q_1 \subset \mathscr{C}(\bar{B}_1),$$

---

[3] This restriction is removed in the remark on p. 212.

[4] Here we tacitly assume that $E \neq X$. If $E = X$, we apply the same argument to a proper subset $E^{(1)} \subset E$ and its complement $E^{(2)}$, eventually obtaining

$$\mu(E^{(1)} - Q^{(1)}) + \mu(E^{(2)} - Q^{(2)}) = \mu(E - Q) < 2\varepsilon,$$

where $Q = Q^{(1)} \cup Q^{(2)}$ and $\Phi(Q) \geqslant c\mu(Q)$ [note that $Q^{(1)}$ and $Q^{(2)}$, like $E^{(1)}$ and $E^{(2)}$, are disjoint].

[5] Here the dummy index α ranges over a set which is general uncountable.

and moreover,

$$\Phi(Q_1) = \sum_{j=1}^{\infty} \Phi(A_j^{(1)}) > c \sum_{j=1}^{\infty} \mu(A_j^{(1)}) = c\mu(Q_1).$$

Next, suppose we have already managed to construct the sets

$$Q_1 \supset \cdots \supset Q_{n-1} = \bigcup_{j=1}^{\infty} A_j^{(n-1)},$$

and the corresponding sets $Z_1, \ldots, Z_{n-1}$ of measure zero. Then the sets Q_n and Z_n can be found as follows: Let

$$Y_n = \bigcup_{j=1}^{\infty} \Gamma(A_j^{(n-1)}) A_j^{n-1},$$

where $\mu(Y_n) = 0$ since every $\Gamma(A_j^{(n-1)})$ is of measure zero. For every point $x \in E - BE - Z_{n-1} - Y_n$ we find all the Vitali sets $A_\alpha^{(n)}(x)$ which contain x, are contained in Q_{n-1}, do not intersect $\bar{B}_1 \cup \cdots \cup \bar{B}_n$ and satisfy the inequality

$$\frac{\Phi[A_\alpha^{(n)}(x)]}{\mu[A_\alpha^{(n)}(x)]} > c.$$

By Properties 2a and 2b, such sets exist and have arbitrarily small measure. Using Property 3, we construct a countable union Q_n of disjoint Vitali sets $A_j^{(n)}$ covering $E - BE - Z_{n-1} - Y_n$ to within a set Y_n' of measure zero. Then

$$E - BE - Z_{n-1} - Y_n - Y_n' \subset Q_n \subset \mathscr{C}(\bar{B}_1 \cup \cdots \cup \bar{B}_n),$$

which can also be written as

$$E - BE - Z_n \subset Q_n \subset \mathscr{C}(\bar{B}_1 \cup \cdots \cup \bar{B}_n),$$

where $Z_n = Z_{n-1} \cup Y_n \cup Y_n'$ is again a set of measure zero. Moreover,

$$\Phi(Q_n) = \sum_{j=1}^{\infty} \Phi(A_j^{(n)}) > c \sum_{j=1}^{\infty} \mu(A_j^{(n)}) = c\mu(Q_n),$$

and we have succeeded in constructing the desired sequence $Q_1 \supset Q_2 \supset \cdots$.

Finally let

$$Q = \bigcap_{n=1}^{\infty} Q_n.$$

Then

$$\Phi(Q) = \lim_{n\to\infty} \Phi(Q_n) \geqslant c \lim_{n\to\infty} \mu(Q_n) = c\mu(Q),$$

while on the other hand,

$$E - BE - \bigcup_{n=1}^{\infty} Z_n \subset Q \subset \mathscr{C}\left(\bigcup_{n=1}^{\infty} B_n\right) \subset E,$$

where

$$\mu(E - Q) \leqslant \mu(BE) + \sum_{n=1}^{\infty} \mu(Z_n) = \mu(B) - \mu(\mathscr{C}E) < \varepsilon.$$

This completes the proof.

Remark. The assumption that X is a Borel set has been used (tacitly) to deduce that $\mathscr{C}E \in \mathfrak{A}$ (where $\mathfrak{A}$ is the underlying σ-ring) and to write $\mu(\mathscr{C}E)$. However, this assumption can be dropped. In fact, if $X \notin \mathfrak{A}$, then $X = X_1 \cup X_2 \cup \cdots$, where $X_1 \subset X_2 \subset \cdots$ and every $X_n \in \mathfrak{A}$. Given any $E \in \mathfrak{A}$, let $E^{(n)} = EX_n$, so that

$$E = \bigcup_{n=1}^{\infty} E^{(n)}.$$

Then, given any $\varepsilon > 0$, we use Lemma 1 to find a set $Q^{(n)} \subset E^{(n)}$ such that

$$\mu(E^{(n)} - Q^{(n)}) < \varepsilon, \qquad \Phi(Q^{(n)}) \geqslant c\mu(Q^{(n)})$$

for every $n = 1, 2, \ldots$, and taking the limit as $n \to \infty$, we obtain

$$\mu(E - Q) \leqslant \varepsilon, \quad \Phi(Q) \geqslant c\mu(Q), \quad Q = \bigcup_{n=1}^{\infty} Q^{(n)},$$

as required.

LEMMA 1′. *If the inequality*

$$\underline{D}\Phi(x) = \varliminf_{\varepsilon \to 0} \frac{\Phi[A_\varepsilon(x)]}{\mu[A_\varepsilon(x)]} < c$$

(c fixed) holds at every point x of a set $E \subset X$ of positive measure, then, given any $\varepsilon > 0$, there is a set $Q \subset E$ such that

$$\mu(E - Q) < \varepsilon, \qquad \Phi(Q) \leqslant c\mu(Q).$$

Proof. Replace $\Phi(E)$ by $2c\mu(E) - \Phi(E)$ in Lemma 1, and use the remark.

LEMMA 2. *The set*

$$E_c = \{x : \bar{D}\Phi(x) \geqslant c\}$$

is measurable for arbitrary real c.

Proof. Given any $\varepsilon > 0$, we can cover every point $x \in E_c$ by a Vitali set $A_\varepsilon(x)$ of measure less than ε such that

$$\frac{\Phi[A_\varepsilon(x)]}{\mu[A_\varepsilon(x)]} > c - \varepsilon.$$

Then the set

$$Q_\varepsilon = \bigcup_{x \in E_c} A_\varepsilon(x)$$

is measurable. To see this, let $\mathfrak{B}$ be the subsystem of *all* Vitali sets with the property that every set of $\mathfrak{B}$ is contained in at least one of the sets $A_\varepsilon(x)$.[6] Then $\mathfrak{B}$ covers Q_ε, and in fact Q_ε coincides with the union of all the sets of $\mathfrak{B}$. It follows from Property 3 that

$$Q_\varepsilon = \bigcup_{j=1}^{\infty} A_j, \qquad A_j \in \mathfrak{B},$$

to within a set of measure zero, and hence Q_ε is measurable, as asserted.

Now let $\{\varepsilon_n\}$ be a sequence of positive numbers approaching zero as $n \to \infty$. Then the set

$$Q = \bigcap_{n=1}^{\infty} Q_{\varepsilon_n}$$

is measurable. Obviously

$$Q \supset E_c, \tag{5}$$

and moreover

$$Q \subset E_c. \tag{6}$$

In fact, given any point $x_0 \in Q$ and any $\varepsilon > 0$, there is at least one set $A_{\varepsilon_n}(x_0) \subset Q_{\varepsilon_n}$ of measure less than ε_n containing x_0, since x_0 also belongs to every Q_{ε_n}. But then

$$\bar{D}\Phi(x_0) = \overline{\lim_{n\to\infty}} \frac{\Phi[A_{\varepsilon_n}(x_0)]}{\mu[A_{\varepsilon_n}(x_0)]} \geqslant \overline{\lim_{n\to\infty}} (c - \varepsilon_n) = c,$$

which proves (6). Comparing (5) and (6), we find that $E_c = Q$. Therefore E_c is measurable, and the lemma is proved.

We are now in a position to prove the basic theorem on differentiation with respect to a Vitali system:

THEOREM 1 (*Lebesgue-Vitali theorem*).[7] *Let X be a set equipped with a measure μ, let $\mathscr{V}$ be a Vitali system of Borel subsets of X, and let $\Phi(E)$ be a countably additive set function on X. Then the derivative of $\Phi(E)$ with respect to $\mathscr{V}$ exists on a set of full measure, and coincides with the derivative of the absolutely continuous component of $\Phi(E)$.*

Proof. Let

$$\Phi(E) = S(E) + A(E), \qquad A(E) = \int_E g(x)\mu(dx), \tag{7}$$

[6] In particular, $\mathfrak{B}$ contains every set $A_\varepsilon(x)$.

[7] The first abstract formulation of the Lebesgue-Vitali theorem (for an absolutely continuous set function) is due to Y. N. Yunovich, *Sur la dérivation des fonctions absolument additives d'ensemble*, Dokl. Akad. Nauk SSSR, **30,** 112 (1941).

where $S(E)$ is singular (but not necessarily continuous), and $A(E)$ is absolutely continuous. We shall prove the theorem in two steps:

Step 1. $DS(x)$ exists almost everywhere and equals zero. Let Z be the set of measure zero on which $S(E)$ is concentrated. Since the positive and negative variations of $S(E)$ are both singular and concentrated on Z, there is no loss of generality in assuming that $S(E) \geqslant 0$ and hence

$$\underline{D}S(x) \geqslant 0. \tag{8}$$

According to Lemma 2, the set

$$E_c = \{x : \bar{D}S(x) \geqslant c\} \cap \mathscr{C}Z$$

is measurable.[8] Moreover, E_c has measure zero, since otherwise, according to Lemma 1, there is a set $Q \subset E_c$ of positive measure such that

$$S(Q) \geqslant \frac{c}{2}\mu(Q) > 0.$$

But this is impossible, since Q does not intersect the set Z on which $S(E)$ is concentrated. Therefore $\mu(E_c) = 0$, and hence

$$\mu(\{x : \bar{D}S(x) > 0\} \cap \mathscr{C}Z = \lim_{c \to 0} \mu(E_c) = 0.$$

It follows that $\bar{D}S(x) = 0$ almost everywhere, and hence, by (8), $DS(x) = 0$ almost everywhere.

Step 2. $DA(x)$ exists almost everywhere and equals $g(x)$. If $\bar{D}A(x) > c$ for all x in a Borel set $X_0 \subset X$, then

$$A(X_0) \geqslant c\mu(X_0). \tag{9}$$

In fact, since the integral in (7) is absolutely continuous (see Property d, p. 124), given any $\varepsilon > 0$, there is a $\delta > 0$ such that $\mu(E) < \delta$ implies $|A(E)| < \varepsilon$. But, according to Lemma 1, we can find a set $Q \subset X_0$ such that

$$\mu(X_0 - Q) < \delta, \qquad A(Q) \geqslant c\mu(Q).$$

It follows that

$$A(X_0) = A(Q) + A(X_0 - Q) > c\mu(Q) - \varepsilon > c\mu(X_0) - \delta - \varepsilon,$$

which proves (9), since δ and ε are arbitrarily small. In the same way, if $\underline{D}A(x) < c$ for all $x \in X_0$, we can use Lemma 1′ to prove that

$$A(X_0) \leqslant c\mu(X_0).$$

Next let

$$E_{ab} = \{x : a < g(x) < b\},$$

---

[8] In writing $\mathscr{C}Z$, we assume that Z is a Borel set, a restriction that can be dropped by using an argument like that given in the remark on p. 212.

where $g(x)$ is the μ-summable function figuring in (7). Then the inequalities

$$\bar{D}A(x) \leqslant b, \qquad \underline{D}A(x) \geqslant a$$

hold almost everywhere on E_{ab}. In fact, suppose $\bar{D}A(x) > b$ on a set $E \subset E_{ab}$ of positive measure. Then, according to (9),

$$A(E) \geqslant b\mu(E),$$

which is impossible, since

$$A(E) = \int_E g(x)\mu(dx) < b\mu(E).$$

Similarly, $\underline{D}A(x) < a$ cannot hold on a set $E \subset E_{ab}$ of positive measure.

Finally, consider the family of all sets of the form

$$E_{r_n s_n} = \{x : r_n < g(x) < s_n\},$$

where r_n and s_n ($r_n < s_n$) are arbitrary rational numbers. As just shown,

$$r_n \leqslant \underline{D}A(x) \leqslant \bar{D}A(x) \leqslant s_n \tag{10}$$

on $E_{r_n s_n}$, except on a set $Z_{r_n s_n}$ of measure zero. It follows that $DA(x)$ exists and equals $g(x)$ everywhere on the set of full measure

$$X' = X - \left(\bigcup_{(r_n, s_n)} Z_{r_n s_n}\right) - Z',$$

where Z' is the set (possibly empty) of measure zero where $g(x)$ takes infinite values. In fact, if $x \in X'$, then (10) holds for every pair of rational numbers r_n and s_n such that

$$r_n < g(x) < s_n.$$

Choosing r_n and s_n arbitrarily close together, we find that

$$\underline{D}A(x) = \bar{D}A(x) = DA(x) = g(x),$$

and the theorem is proved.

10.4. Some Consequences of the Lebesgue-Vitali Theorem

We now examine some of the implications of Theorem 1, first for the case of nets and then for a family of cubes.

10.4.1. De Possel's theorem. We begin by proving

THEOREM 2. *Let X be a set equipped with a measure μ, such that every set $\{x\}$ consisting of a single point $x \in X$ is measurable, with measure zero. Then every net $\mathfrak{N}$ of subsets of X is a Vitali system.*

Proof. By definition, $\mathfrak{N}$ is completely sufficient, which verifies Property 1, p. 209. To verify Property 2a, we choose the empty set $\varnothing$ as the boundary $\Gamma(E)$ of every set $E \in \mathfrak{N}$. Then, given any point $x \in E$, $E \in \mathfrak{N}$, let E' be any other set in $\mathfrak{N}$ containing x. Since two sets E, $E' \in \mathfrak{N}$ are either disjoint (in particular, this is the case if E and E' are of the same rank) or else one of the sets is a proper subset of the other, it follows that every set in $\mathfrak{N}$ of sufficiently small measure containing x is contained in E. As for Property 2b, we need only note that if x does not belong to a given set $E \in \mathfrak{N}$, then x must belong to some other set $E' \in \mathfrak{N}$, $EE' = \varnothing$. Therefore every set $A \in \mathfrak{N}$ of sufficiently small measure containing x belongs to E', i.e., cannot intersect E. Finally, to prove Property 3, suppose $E \subset X$ is a set covered by a subfamily $\mathfrak{B} \subset \mathfrak{N}$. Then we can always eliminate intersecting sets, by the simple device of first choosing all sets of the first rank in $\mathfrak{B}$, then all the sets of the second rank in $\mathfrak{B}$ not contained in those already chosen, and so on. This gives a sequence of disjoint sets A_j $(j = 1, 2, \ldots)$ covering E, as required.[9]

COROLLARY 1 (*De Possel's theorem*). *Let X and $\mathfrak{N}$ be the same as in Theorem 2, and let $\Phi(E)$ be a countably additive set function on X. Then the derivative of $\Phi(E)$ with respect to $\mathfrak{N}$ exists on a set of full measure, and coincides with the density of the absolutely continuous component of $\Phi(E)$.*

Proof. Apply the Lebesgue-Vitali theorem.

COROLLARY 2. *If $F(x)$ is a function of bounded variation in $[a, b]$, then the derivative of $F(x)$ in the sense of formula* (2) *exists almost everywhere and equals the density of the absolutely continuous component of $F(x)$.*

10.4.2. Lebesgue's theorem on differentiation of a function of bounded variation. Next we consider Vitali systems of a particularly important kind:

THEOREM 3. *Let X be a bounded n-dimensional basic block* **B**, *equipped with ordinary Lebesgue measure. Then the family $\mathfrak{Q}$ of all closed cubes of the form*

$$Q = \{x : a_1 \leqslant x_1 \leqslant a_1 + h, \ldots, a_n \leqslant x_n \leqslant a_n + h\} \subset \mathbf{B}$$

($n > 0$) *is a Vitali system.*

Proof. It is clear that Properties 1, 2a and 2b, p. 209 are satisfied if the boundary $\Gamma(Q)$ of every cube $Q \in \mathfrak{Q}$ is defined to be its ordinary topological boundary (i.e., the intersection of the closure of Q with the closure of its complement).[10] To verify Property 3, we use the following

---

[9] Note that any net $\mathfrak{N}$ (and hence any subset of $\mathfrak{N}$) is countable.

[10] Then $Q - Q\Gamma(Q)$ is the interior of Q and $\bar{Q} = Q \cup \Gamma(Q)$ the closure of Q (as anticipated by the notation).

proof (due to Banach): Let $E \subset \mathbf{B}$ be a set covered by a subsystem $\mathfrak{B} \subset \mathfrak{Q}$ of cubes such that for any $x \in E$ and any $\varepsilon > 0$, there is a cube $A_\varepsilon(x) \in \mathfrak{B}$ with volume $s[A_\varepsilon(x)]$ less than ε which contains x. Writing

$$k_1 = \sup_{A_\alpha \in \mathfrak{B}_1} s(A_\alpha),$$

where $\mathfrak{B}_1 \equiv \mathfrak{B}$, we choose any cube Q_1 in $\mathfrak{B}_1$ with volume greater than $k_1/2$. If Q_1 does not cover E to within a set of measure zero, then the set $\mathfrak{B}_2 \subset \mathfrak{B}_1$ of cubes disjoint from Q_1 is nonempty. In this case, we write

$$k_2 = \sup_{A_\alpha \in \mathfrak{B}_2} s(A_\alpha),$$

and choose any cube $Q_2 \in \mathfrak{B}_2$ with volume greater than $k_2/2$. Continuing this process, we either eventually manage to cover E (to within a set of measure zero) by finitely many cubes in $\mathfrak{B}$, thereby establishing Property 3 at once, or else we obtain sets $\mathfrak{B} \equiv \mathfrak{B}_1 \supset \mathfrak{B}_2 \supset \cdots$, numbers $k_1 \geqslant k_2 \geqslant \cdots$ and disjoint cubes $Q_1, Q_2, \ldots$ such that

$$Q_j \in \mathfrak{B}_j, \quad s(Q_j) > \frac{k_j}{2} \qquad (j = 1, 2, \ldots).$$

As we now show, the sequence $\{Q_j\}$ covers E to within a set of measure zero, so that Property 3 holds in any event.

In fact, let Z be the subset of E such that $ZQ_j = \varnothing$ $(j = 1, 2, \ldots)$. Then, given any integer $p > 1$ and any point $x_0 \in Z$, we have

$$x_0 \notin Q_j \qquad (j = 1, \ldots, p-1).$$

On the other hand, since $x_0 \in E$, there are cubes in $\mathfrak{B}$ of arbitrarily small volume containing x_0. In particular, there is a cube $Q \in \mathfrak{B}$ containing x_0 such that $QQ_1 = \cdots = QQ_{p-1} = \varnothing$, and hence $Q \in \mathfrak{B}_p$. If moreover $QQ_p = \cdots = QQ_{r-1} = \varnothing$, then Q belongs to $\mathfrak{B}_{p+1}, \ldots, \mathfrak{B}_r$, and hence $s(Q) \leqslant k_r$. But the series

$$\sum_{j=1}^{\infty} k_j \leqslant 2 \sum_{j=1}^{\infty} s(Q_j) \leqslant 2s(\mathbf{B})$$

converges, so that $k_j \to 0$ as $j \to \infty$. Thus there is a first index $r \geqslant p$ such that Q intersects Q_r. Let l denote the side length of Q and l_r that of Q_r. Since $Q \in \mathfrak{B}_r$, we have $s(Q) \leqslant k_r$, i.e., $l \leqslant \sqrt[n]{k_r}$. Moreover $s(Q_r) > k_r/2$, by construction, and hence $l_r > \sqrt[n]{k_r/2}$. Since Q intersects Q_r, Q is contained in the cube with the same center as Q_r but with side length

$$l_r + 2l \leqslant l_r + 2\sqrt[n]{k_r} < l_r + 2\sqrt[n]{2}\,l_r = (1 + 2\sqrt[n]{2})l_r \leqslant 5l_r,$$

and hence certainly in the cube $\hat{Q}_r$ with the same center as Q_r but with side length five times that of Q_r. It follows that

$$x_0 \in \hat{Q}_r \subset \bigcup_{j=p}^{\infty} \hat{Q}_j,$$

and hence

$$Z \subset \bigcup_{j=p}^{\infty} \hat{Q}_j$$

for every $p > 1$, since x_0 is an arbitrary point of Z. But $s(\hat{Q}_j) = 5^n s(Q_j) \to 0$ as $j \to \infty$, and hence Z is a set of measure zero. This completes the proof.

COROLLARY 1. *Let $\Phi(E)$ be a countably additive set function on a bounded n-dimensional block* **B**, *and let $x_0 = (x_1^{(0)}, \ldots, x_n^{(0)})$ be a point of* **B**. *Then the derivative in the sense*

$$D\Phi(x) = \lim_{h \to 0} \frac{\Phi[Q_h(x_0)]}{s[Q_h(x_0)]},$$

where $Q_h(x_0)$ is the cube[11]

$$\{x\colon x_1^{(0)} \leqslant x_1 \leqslant x_1^{(0)} + h, \ldots, x_n^{(0)} \leqslant x_n \leqslant x_n^{(0)} + h\},$$

exists almost everywhere and coincides with the density of the absolutely continuous component of $\Phi(E)$.

COROLLARY 2 (*Lebesgue's theorem on differentiation of a function of bounded variation*). *If $F(x)$ is of bounded variation in $[a, b]$, then the "ordinary" derivative*

$$F'(x) = \lim_{h \to 0} \frac{F(x+h) - F(x)}{h}$$

exists almost everywhere and equals the density of the absolutely continuous component of $F(x)$.

Proof. If $\Phi(E)$ is the set function with generating function $F(x)$, then the difference between the quantity $\Phi[x, x+h] = \Phi(x, x+h] + \Phi(\{x\})$ figuring in Corollary 1 and the quantity $\Phi(x, x+h] = F(x+h) - F(x)$ does not matter, since $\Phi(\{x\}) \neq 0$ for at most countably many points x (why?) and hence $\Phi(\{x\}) = 0$ on a set of full measure.

---

[11] Here, as on p. 206, we allow $h < 0$ by changing $Q_h(x_0)$ to

$$\{x\colon x_1^{(0)} + h \leqslant x_1 \leqslant x_1^{(0)}, \ldots, x_n^{(0)} + h \leqslant x_n \leqslant x_n^{(0)}\}$$

and $\Phi[Q_h(x_0)]$ to $-\Phi[Q_h(x_0)]$.

COROLLARY 3. *If $F(x)$ is absolutely continuous in $[a, b]$, then $F'(x)$ exists almost everywhere. Moreover, $F'(x)$ is summable and*[12]

$$F(x) - F(a) = \int_a^x F'(\xi)\, d\xi. \tag{11}$$

COROLLARY 4. *If $F(x)$ is singular in $[a, b]$, then $F'(x) = 0$ almost everywhere.*

COROLLARY 5. *If $F(x)$ is of bounded variation in $[a, b]$ and if $F'(x) = 0$ almost everywhere, then $F(x)$ is singular in $[a, b]$.*

Proof. Writing $F(x)$ in the form

$$F(x) = A(x) + S(x),$$

where $A(x)$ is absolutely continuous and $S(x)$ is singular, we use Corollary 4 to deduce that $A'(x) = 0$ almost everywhere. But then the absolutely continuous set function

$$A(E) = \int_E A'(x)\, dx$$

vanishes on every Borel set E, and hence $A(x) = A[a, x] \equiv 0$, i.e., $F(x) \equiv S(x)$, as asserted.

Remark 1. As shown in Prob. 3, p. 223, there exist functions whose derivatives vanish almost everywhere but which are not of bounded variation (and hence not singular).

Remark 2. Let $F(x)$ be a generating function of bounded variation in $[a, b]$, with decomposition

$$F(x) = A(x) + S(x),$$

where $A(x)$ is absolutely continuous and $S(x)$ is singular. Then $A(x)$ and $S(x)$ can be found from $F(x)$ itself by using the formulas

$$A(x) = \int_a^x F'(\xi)\, d\xi, \qquad S(x) = F(x) - \int_a^x F'(\xi)\, d\xi.$$

THEOREM 4. *If $F(x)$ is a generating function of bounded variation in $[a, b]$, with positive variation $P(x)$, negative variation $Q(x)$ and total variation $V(x)$, then*

$$P'(x) = [F'(x)]^+, \qquad Q'(x) = [F'(x)]^-, \qquad V'(x) = |F'(x)| \tag{12}$$

almost everywhere. Moreover,

$$P(x) = \int_a^x [F'(\xi)]^+\, d\xi, \quad Q(x) = \int_a^x [F'(\xi)]^-\, d\xi, \quad V(x) = \int_a^x |F'(\xi)|\, d\xi, \tag{13}$$

if $F(x)$ is absolutely continuous in $[a, b]$.

---

[12] Formula (11) is a far-reaching generalization of the classical relation between definite integrals and primitives.

Proof. Let $\Phi(E)$ be the set function with generating function $F(x)$, and let X^+ and X^- be the two sets figuring in the corresponding Hahn decomposition of $X = [a, b]$ (see Theorem 10, p. 163). Then $\Phi(E) \geqslant 0$ on every Borel set $E \subset X^+$. But then $F'(x) \geqslant 0$ almost everywhere on X^+, since otherwise $F'(x) \leqslant -c < 0$ on a set $E_0 \subset X^+$ of positive measure, and then according to Lemma 1′, p. 212, there would be a set $Q \subset E_0$ of positive measure such that $\Phi(Q) \leqslant -c\mu(Q)$, which is impossible. Similarly, $F'(x) \leqslant 0$ almost everywhere on X^-.

Next let

$$\Phi(E) = p(E) - q(E)$$

be the representation of $\Phi(E)$ in terms of its positive variation $p(E)$ and negative variation $q(E)$ [see Theorem 8, p. 160]. Then $P(x)$ is the generating function of $p(E)$ and $Q(x)$ the generating function of $q(E)$. Since $q(E)$ vanishes on X^+ and all its Borel subsets, we have $Q'(x) = 0$ almost everywhere on X^+, by substantially the same argument used to prove that $F'(x) \geqslant 0$ on X^+, and similarly $P'(x) = 0$ almost everywhere on X^-. Since $F'(x) = P'(x) - Q'(x)$, it follows that $F'(x) = P'(x)$ almost everywhere on X^+ and $F'(x) = -Q'(x)$ almost everywhere on X^-. Together with the formula $V'(x) = P'(x) - Q'(x)$ and the properties of $F'(x)$ just proved, this implies (12). Moreover, if $F(x)$ is absolutely continuous in $[a, b]$, then so are the functions $P(x)$, $Q(x)$ and $V(x)$ [see p. 184,] which therefore equal the integrals of their own derivatives, by Corollary 3, thereby proving (13).

10.5. Differentiation with Respect to the Underlying σ-Ring

To avoid repetition, we establish the convention that throughout this section *X is a set equipped with a σ-ring $\mathfrak{A}$ of Borel sets $E \subset X$, a Borel measure μ, a summable function $\varphi(x)$ and a Vitali system $\mathscr{V}$ of sets $E \in \mathfrak{A}$.*

DEFINITION 1. *A point $x_0 \in X$ is said to be a Lebesgue point of $\varphi(x)$ (relative to $\mathscr{V}$) if*

$$\lim_{\varepsilon \to 0} \frac{1}{\mu[A_\varepsilon(x_0)]} \int_{A_\varepsilon(x_0)} |\varphi(x) - \varphi(x_0)|\, \mu(dx) = 0, \tag{14}$$

where $A_\varepsilon(x_0)$ is any Vitali set of measure less than ε containing x_0.

THEOREM 5. *Almost every point of X is a Lebesgue point of $\varphi(x)$.*

Proof. Given any real number r, it follows from the Lebesgue-Vitali theorem that

$$\lim_{\varepsilon \to 0} \frac{1}{\mu[A_\varepsilon(x_0)]} \int_{A_\varepsilon(x_0)} |\varphi(x) - r|\, \mu(dx) = |\varphi(x_0) - r| \tag{15}$$

for all x_0 in a set E_r of full measure. Then the set

$$E = \bigcap_{n=1}^{\infty} E_{r_n},$$

where r_n ranges over all the rational numbers, is also of full measure. Let x_0 be a point of E such that $\varphi(x_0)$ is finite. Then, given any $\delta > 0$, there is a rational number such that $|\varphi(x_0) - r| < \delta/3$, and hence

$$\frac{1}{\mu[A_\varepsilon(x_0)]} \int_{A_\varepsilon(x_0)} |\varphi(x) - \varphi(x_0)| \, \mu(dx)$$

$$\leqslant \frac{1}{\mu[A_\varepsilon(x_0)]} \int_{A_\varepsilon(x_0)} |\varphi(x) - r| \, \mu(dx) + \frac{1}{\mu[A_\varepsilon(x_0)]} \int_{A_\varepsilon(x_0)} |r - \varphi(x_0)| \, \mu(dx)$$

$$= \left\{\frac{1}{\mu[A_\varepsilon(x_0)]} \int_{A_\varepsilon(x_0)} |\varphi(x) - r| \, \mu(dx) - |\varphi(x_0) - r|\right\} + 2\,|\varphi(x_0) - r|\,. \tag{16}$$

Because of (15), the term in braces can be made less than $\delta/3$ for sufficiently small ε. Therefore the left-hand side of (16) can be made less than δ for sufficiently small ε. This proves the theorem, since $\varphi(x)$ is finite almost everywhere.

Next we generalize the notion of "regular convergence," already encountered on p. 207:

DEFINITION 2. *A sequence $E_1, E_2, \ldots$ of Borel sets is said to converge regularly to a point $x_0 \in X$ if*

1) *Every E_n is contained in a Vitali set A_n such that $x_0 \in A_n$ and $\mu(A_n) \to 0$ as $n \to \infty$;*
2) *There is a fixed constant $c > 0$ such that*

$$\mu(E_n) \geqslant c\mu(A_n)$$

for every n.

We are now in a position to generalize differentiation in the sense of formula (3), p. 207:

DEFINITION 3. *Let $\Phi(E)$ be a countably additive set function defined on the σ-ring $\mathfrak{A}$ (and hence on the Vitali system $\mathscr{V}$). Then by the derivative of $\Phi(E)$ at the point $x_0 \in X$ with respect to $\mathfrak{A}$ is meant the quantity*

$$D_{\mathfrak{A}}\Phi(x_0) \equiv \lim_{n\to\infty} \frac{\Phi(E_n)}{\mu(E_n)}$$

(provided the limit exists), where $E_1, E_2, \ldots$ is any sequence of Borel sets regularly converging to x_0.

THEOREM 6. *If $E_1, E_2, \ldots$ is a sequence of Borel sets converging regularly to a Lebesgue point x_0 of $\varphi(x)$, then*

$$\lim_{n\to\infty} \frac{1}{\mu(E_n)}\int_{E_n} \varphi(x)\mu(dx) = \varphi(x_0).$$

Proof. We need only note that

$$\begin{aligned}\left|\varphi(x_0) - \frac{1}{\mu(E_n)}\int_{E_n} \varphi(x)\mu(dx)\right| &= \left|\frac{1}{\mu(E_n)}\int_{E_n} [\varphi(x_0) - \varphi(x)]\mu(dx)\right| \\ &\leqslant \frac{1}{\mu(E_n)}\int_{E_n} |\varphi(x) - \varphi(x_0)|\, \mu(dx) \\ &\leqslant \frac{1}{c\mu(A_n)}\int_{A_n} |\varphi(x) - \varphi(x_0)|\, \mu(dx) \to 0 \quad \text{as} \quad n \to \infty.\end{aligned}$$

COROLLARY. *Let*

$$\Phi(E) = \int_E \varphi(x)\mu(dx)$$

be the "indefinite integral" of $\varphi(x)$. Then

$$D_{\mathfrak{A}}\Phi(x_0) = \varphi(x_0)$$

at every Lebesgue point x_0 of $\varphi(x)$.

THEOREM 7. *If $\Phi(E)$ is a countably additive set function on X, then the derivative of $\Phi(E)$ with respect to the underlying σ-ring $\mathfrak{A}$ exists on a set of full measure, and coincides with the density of the absolutely continuous component of $\Phi(E)$.*

Proof. If $\Phi(E)$ is absolutely continuous, the result follows at once from Theorem 5 and the corollary to Theorem 6. Thus suppose $\Phi(E)$ is singular and nonnegative (the latter assumption entails no loss of generality), and let x_0 be a point at which the derivative $D_{\mathscr{V}}\Phi(E)$ with respect to the Vitali system $\mathscr{V}$ exists and equals zero. As we know from Theorem 1, such points form a set of full measure. If $E_1, E_2, \ldots$ is a sequence of Borel sets converging regularly to x_0, then $\mu(E_n) \geqslant c\mu(A_n)$ and hence

$$D_{\mathfrak{A}}\Phi(x_0) = \lim_{n\to\infty} \frac{\Phi(E_n)}{\mu(E_n)} \leqslant \lim_{n\to\infty} \frac{1}{c}\frac{\Phi(A_n)}{\mu(A_n)} = \frac{1}{c} D_{\mathscr{V}}\Phi(x_0) = 0,$$

where the A_n are suitable Vitali sets. The theorem now follows from the observation that a general set function $\Phi(E)$ is the sum of its absolutely continuous and singular components.[13]

---

[13] It should be kept in mind that the singular component of $\Phi(E)$, unlike its absolutely continuous component, may not be defined on all Lebesgue-measurable subsets $E \subset X$ (recall Prob. 10, p. 204).

COROLLARY. *If $F(x) = \Phi[a, x]$ is of bounded variation in $[a, b]$, then the derivative in the sense*

$$F'(x_0) = \lim_{n\to\infty} \frac{\Phi(E_n)}{\mu(E_n)},$$

where $E_1, E_2, \ldots$ is a sequence of Borel sets converging regularly to x_0,[14] *exists almost everywhere and equals the absolutely continuous component of $F(x)$.*

PROBLEMS

1. Construct a function $F(x)$ with Properties a and b listed on p. 207.

Hint. As $n \to \infty$ the right-hand end points of the intervals $p2^{-n} < x \leqslant (p + 1)2^{-n}$ containing x_0 form a strictly decreasing sequence $\{x_n\}$ converging to x_0, where obviously

$$x_{n+1} - x_0 < \frac{x_n - x_0}{2}.$$

Let $F(x)$ be a continuous function equal to zero at the points x_m and linear in the intervals (x_{n+1}, ξ_n), (ξ_n, x_n) where

$$\xi_n = \frac{x_n + x_{n+1}}{2},$$

and let $F(x)$ take the value $\xi_n - x_0$ at the midpoint ξ_n. Then $F(x)$ clearly has no derivative in the ordinary sense at x_0. Verify that $F(x)$ is of bounded variation.

2. Show that the function

$$F(x) = x^2 \sin \frac{1}{x}$$

fails to have a derivative in the sense of formula (3), p. 207 at the point $x_0 = 0$, although its ordinary derivative exists and equals zero at x_0.

Hint. Choose E_n to be the set of intervals

$$(\alpha_k^{(n)}, \beta_k^{(n)}] \subset \left(-\frac{1}{n}, \frac{1}{n}\right]$$

on which $F(x)$ increases.

3. The function

$$F(x) = x \sin \frac{1}{x}$$

---

[14] In particular, $E_1, E_2, \ldots$ can be any sequence of Borel sets such that 1) every E_n is contained in the interval $(x_0 - h_n, x_0 + h_n]$, where $h_n \to 0$ as $n \to \infty$, and 2) there is a fixed constant $c > 0$ such that $\mu(E_n) \geqslant 2ch_n$ for every n (cf. p. 207).

is not of bounded variation in $[-\pi, \pi]$. By replacing $F(x)$ on its intervals of monotonicity by a function of the Cantor type (see Prob. 2, p. 86), construct a continuous function whose derivative vanishes almost everywhere but which is not of bounded variation (and hence not singular).

4 (*Fubini's convergence theorem*). Given a σ-ring $\mathfrak{A}$, let

$$\sum_{n=1}^{\infty} \Phi_n(E)$$

be a series of nonnegative countably additive set functions converging on every set $E \in \mathfrak{A}$ to a finite function $\Phi(E)$, which is itself countably additive by Prob. 5, p. 203. Prove that

$$\sum_{n=1}^{\infty} D\Phi_n(x) = D\Phi(x)$$

almost everywhere.

Hint. On the set of full measure where all the derivatives $D\Phi(x)$, $D\Phi_1(x)$, $D\Phi_2(x), \ldots$ exist, take the limit of the inequality

$$\sum_{n=1}^{N} \frac{\Phi_n[A(x_0)]}{\mu[A(x_0)]} \leqslant \frac{\Phi[A(x_0)]}{\mu[A(x_0)]},$$

obtaining

$$\sum_{n=1}^{\infty} D\Phi_n(x) \leqslant D\Phi(x).$$

Given any integer $k > 0$, let N_k be such that

$$\Phi(X) - \sum_{n=1}^{N_k} \Phi_n(X) < \frac{1}{2^k}.$$

Then the series with general term

$$\Psi_k(E) = \Phi(E) - \sum_{n=1}^{N_k} \Phi_k(E)$$

converges for every $E \in \mathfrak{A}$, and the series with general term

$$D\Psi_k(x) = D\Phi(x) - \sum_{n=1}^{N_k} D\Phi_n(x)$$

converges on a set of full measure. Therefore

$$\sum_{k=1}^{N_k} D\Phi_n(x) \to D\Phi(x),$$

and hence

$$\sum_{n=1}^{N} D\Phi_n(x) \to D\Phi(x).$$

5. Given a summable nonnegative function $g(x)$ defined on a set X equipped with a Lebesgue measure μ, define a new countably additive set function $G(E) = I(g\chi_E)$. Regarding $G(E)$ as a new elementary measure, extend $G(E)$ to a new

Lebesgue measure, and construct the corresponding space of summable functions L_G. Show that a function $\varphi(x)$ belongs to L_G if and only if φg belongs to L, and moreover

$$I_G\varphi = I(\varphi g).$$

Hint. Generalize the considerations of Example 2, pp. 90, 145.

6.[15] Let X be an arbitrary set equipped with a measure μ, and let $\mathfrak{B}$ be a family of Borel subsets of X which covers a given set $E \subset X$ and has the following properties:

1) Given any $x \in E$ and any $\varepsilon > 0$, there is a set in $\mathfrak{B}$ of measure less than ε which contains x;
2) If $x \notin A_0$, $A_0 \in \mathfrak{B}$, then every set in $\mathfrak{B}$ of sufficiently small measure containing x does not intersect A_0;
3) There exist two positive numbers a and b such that

$$\mu\left(\bigcup_{\substack{\mu(A) \leqslant a \\ AA_0 \neq \varnothing}} A\right) \leqslant b\mu(A_0).$$

Prove that E can be covered to within a set of measure zero by countably many disjoint sets $A_j \in \mathfrak{B}$.

Hint. The proof resembles that of Theorem 3, p. 216, but use of Property 3 replaces the argument involving the side lengths of the cubes Q and Q_r.

7. Let X and Y be two sets, equipped with (nonnegative) measures μ and ν, respectively, and suppose $y = f(x)$ is a one-to-one mapping of X onto Y, or at least "almost one-to-one" (i.e., one-to-one after deleting a set of μ-measure zero from X and a set of ν-measure zero from Y). Moreover, suppose the mapping $y = f(x)$ carries measurable sets into measurable sets and sets of μ-measure zero into sets of ν-measure zero. Let $\mathscr{V}$ be a Vitali system of subsets of X, and define the "Jacobian of the transformation" by the formula

$$\mathscr{D}\begin{pmatrix} y \\ x \end{pmatrix} = \lim_{\varepsilon \to 0} \frac{f[A_\varepsilon(x_0)]}{\mu[A_\varepsilon(x_0)]}, \tag{17}$$

where $A_\varepsilon(x_0)$ is any Vitali set of measure less than ε containing the point x_0. Show that the function (17) exists almost everywhere on X and is summable on every summable set $E \subset X$.

Hint. The function $\Phi(E) = \nu[f(E)]$ is countably additive and absolutely continuous.

---

[15] Due to V. A. Yashnikov.

8. With the same notation as in the preceding problem, prove the validity of of the formula for "changing variables"

$$\int_Y \varphi(y)\nu(dy) = \int_X \varphi[f(x)]\mathscr{D}\binom{y}{x}\,\mu(dx),$$

where the existence of either side implies that of the other.

Hint. Introduce the new measure

$$\Phi(E) = \nu[f(E)] = \int_E \mathscr{D}\binom{y}{x}\,\mu(dx)$$

on all summable subsets $E \subset X$, and then use Prob. 5, p. 224.

BIBLIOGRAPHY

Berberian, S. K., *Measure and Integration*, The Macmillan Co., New York (1965).

Burkill, J. C., *The Lebesgue Integral*, Cambridge University Press, London (1953).

Dunford, N. and J. T. Schwartz, *Linear Operators, Part I: General Theory*, Interscience Publishers, Inc., New York (1958).

Goffman, C., *Real Functions*, Holt, Rinehart and Winston, Inc., New York (1953).

Graves, L. M., *The Theory of Functions of Real Variables*, second edition, McGraw-Hill Book Co., New York (1956).

Hahn, H. and A. Rosenthal, *Set Functions*, University of New Mexico Press, Albuquerque, New Mexico (1948).

Halmos, P. R., *Measure Theory*, D. Van Nostrand Co., Inc., Princeton, N.J. (1950).

Hartman, S. and J. Mikusiński, *The Theory of Lebesgue Measure and Integration* (translated by L. F. Boron), Pergamon Press, Oxford (1961).

Hewitt, E. and K. Stromberg, *Real and Abstract Analysis*, Springer-Verlag, Inc., New York (1965).

Hildebrandt, T. H., *Introduction to the Theory of Integration*, Academic Press, Inc., New York (1963).

Jeffery, R. L., *The Theory of Functions of a Real Variable*, second edition, University of Toronto Press, Toronto (1953).

Kestelman, H., *Modern Theories of Integration*, second revised edition, Dover Publications, Inc., New York (1960).

Kolmogorov, A. N. and S. V. Fomin, *Measure, Lebesgue Integrals, and Hilbert Space* (translated by N. A. Brunswick and A. Jeffrey), Academic Press, Inc., New York (1961).

Loomis, L. H., *An Introduction to Abstract Harmonic Analysis*, D. Van Nostrand Co., Inc., Princeton, N.J. (1953).

McShane, E. J., *Integration*, Princeton University Press, Princeton, N.J. (1944).

McShane, E. J. and T. A. Botts, *Real Analysis*, D. Van Nostrand Co., Inc., Princeton, N.J. (1959).

Munroe, M. E., *Introduction to Measure and Integration*, Addison-Wesley Publishing Co., Inc., Reading, Mass. (1953).

Natanson, I. P., *Theory of Functions of a Real Variable* (translated by L. F. Boron, with the collaboration of E. Hewitt), Frederick Ungar Publishing Co., New York. *Volume I* (1955), *Volume II* (1960).

Riesz, F. and B. Sz.-Nagy, *Functional Analysis* (translated by L. F. Boron), Frederick Ungar Publishing Co., New York (1955).

Rogosinski, W., *Volume and Integral*, Interscience Publishers, Inc., New York (1962).

Royden, H. L., *Real Analysis*, The Macmillan Co., New York (1963).

Saks, S., *Theory of the Integral* (translated by L. C. Young, with two notes by S. Banach), second revised edition, Dover Publications, Inc., New York (1964).

Sz.-Nagy, B., *Introduction to Real Functions and Orthogonal Expansions*, Oxford University Press, New York (1965).

Taylor, A. E., *General Theory of Functions and Integration*, Blaisdell Publishing Co., New York (1965).

Titchmarsh, E. C., *The Theory of Functions*, second edition, Oxford University Press, New York (1939).

Von Neumann, J., *Functional Operators, Volume I: Measures and Integrals*, Princeton University Press, Princeton, N.J. (1950).

Williamson, J. H., *Lebesgue Integration*, Holt, Rinehart and Winston, Inc., New York (1962).

Zaanen, A. C., *An Introduction to the Theory of Integration*, Interscience Publishers Inc., New York (1958).

INDEX

J

K

L

M

N

O

P

Q

R

S

T

U

V

W

Y

A CATALOGUE OF SELECTED DOVER BOOKS IN ALL FIELDS OF INTEREST

A CATALOGUE OF SELECTED DOVER BOOKS IN ALL FIELDS OF INTEREST

CREATIVE LITHOGRAPHY AND HOW TO DO IT, Grant Arnold. Lithography as art form: working directly on stone, transfer of drawings, lithotint, mezzotint, color printing; also metal plates. Detailed, thorough. 27 illustrations. 214pp.
21208-4 Pa. $3.00

DESIGN MOTIFS OF ANCIENT MEXICO, Jorge Enciso. Vigorous, powerful ceramic stamp impressions — Maya, Aztec, Toltec, Olmec. Serpents, gods, priests, dancers, etc. 153pp. 6⅛ x 9¼. 20084-1 Pa. $2.50

AMERICAN INDIAN DESIGN AND DECORATION, Leroy Appleton. Full text, plus more than 700 precise drawings of Inca, Maya, Aztec, Pueblo, Plains, NW Coast basketry, sculpture, painting, pottery, sand paintings, metal, etc. 4 plates in color. 279pp. 8⅜ x 11¼. 22704-9 Pa. $4.50

CHINESE LATTICE DESIGNS, Daniel S. Dye. Incredibly beautiful geometric designs: circles, voluted, simple dissections, etc. Inexhaustible source of ideas, motifs. 1239 illustrations. 469pp. 6⅛ x 9¼. 23096-1 Pa. $5.00

JAPANESE DESIGN MOTIFS, Matsuya Co. Mon, or heraldic designs. Over 4000 typical, beautiful designs: birds, animals, flowers, swords, fans, geometric; all beautifully stylized. 213pp. 11⅜ x 8¼. 22874-6 Pa. $4.95

PERSPECTIVE, Jan Vredeman de Vries. 73 perspective plates from 1604 edition; buildings, townscapes, stairways, fantastic scenes. Remarkable for beauty, surrealistic atmosphere; real eye-catchers. Introduction by Adolf Placzek. 74pp. 11⅜ x 8¼. 20186-4 Pa. $2.75

EARLY AMERICAN DESIGN MOTIFS, Suzanne E. Chapman. 497 motifs, designs, from painting on wood, ceramics, appliqué, glassware, samplers, metal work, etc. Florals, landscapes, birds and animals, geometrics, letters, etc. Inexhaustible. Enlarged edition. 138pp. 8⅜ x 11¼. 22985-8 Pa. $3.50
23084-8 Clothbd. $7.95

VICTORIAN STENCILS FOR DESIGN AND DECORATION, edited by E.V. Gillon, Jr. 113 wonderful ornate Victorian pieces from German sources; florals, geometrics; borders, corner pieces; bird motifs, etc. 64pp. 9⅜ x 12¼. 21995-X Pa. $2.50

ART NOUVEAU: AN ANTHOLOGY OF DESIGN AND ILLUSTRATION FROM THE STUDIO, edited by E.V. Gillon, Jr. Graphic arts: book jackets, posters, engravings, illustrations, decorations; Crane, Beardsley, Bradley and many others. Inexhaustible. 92pp. 8⅛ x 11. 22388-4 Pa. $2.50

ORIGINAL ART DECO DESIGNS, William Rowe. First-rate, highly imaginative modern Art Deco frames, borders, compositions, alphabets, florals, insectals, Wurlitzer-types, etc. Much finest modern Art Deco. 80 plates, 8 in color. 8⅜ x 11¼. 22567-4 Pa. $3.00

HANDBOOK OF DESIGNS AND DEVICES, Clarence P. Hornung. Over 1800 basic geometric designs based on circle, triangle, square, scroll, cross, etc. Largest such collection in existence. 261pp. 20125-2 Pa. $2.50

8659